Threads of Labour

ANTIPODE BOOK SERIES

General Editor: Noel Castree, Professor of Geography, University of Manchester, UK
Like its parent journal, the Antipode Book series reflects distinctive new developments in radical geography. It publishes books in a variety of formats – from reference books to works of broad explication to titles that develop and extend the scholarly research base – but the commitment is always the same: to contribute to the praxis of a new and more just society.

Published

David Harvey: A Critical Reader
Edited by Noel Castree and Derek Gregory

Threads of Labour
Edited by Angela Hale and Jane Wills

Life's Work: Geographies of Social Reproduction
Edited by Katharyne Mitchell, Sallie A. Marston and Cindi Katz

Redundant Masculinities? Employment Change and White Working Class Youth
Linda McDowell

Spaces of Neoliberalism
Edited by Neil Brenner and Nik Theodore

Space, Place and the New Labour Internationalism
Edited by Peter Waterman and Jane Wills

Forthcoming

Working the Spaces of Neoliberalism: Activism, Professionalisation and Incorporation
Edited by Nina Laurie and Liz Bondi

Neo-liberalization: Borders, Edges, Frontiers, Peoples
Edited by Kim England & Kevin Ward

Cities of Whiteness
Wendy Shaw

The South Strikes Back: Labour in the Global Economy
Rob Lambert and Edward Webster

Threads of Labour

Garment Industry Supply Chains from the Workers' Perspective

Edited by
Angela Hale and Jane Wills

Women Working Worldwide

Blackwell
Publishing

© 2005 by Blackwell Publishing Ltd

BLACKWELL PUBLISHING
350 Main Street, Malden, MA 02148-5020, USA
9600 Garsington Road, Oxford OX4 2DQ, UK
550 Swanston Street, Carlton, Victoria 3053, Australia

The right of Angela Hale and Jane Wills to be identified as the Authors of the Editorial Material in this Work has been asserted in accordance with the UK Copyright, Designs, and Patents Act 1988.

First published 2005 by Blackwell Publishing Ltd

1 2005

Library of Congress Cataloging-in-Publication Data

Threads of labour : garment industry supply chains from the workers' perspective / edited by Angela Hale and Jane Wills.
     p. cm. — (Antipode book series)
   Includes bibliographical references and index.
   ISBN-13: 978-1-4051-2637-3 (hardback : alk. paper)
   ISBN-10: 1-4051-2637-X (hardback : alk. paper)
   ISBN-13: 978-1-4051-2638-0 (pbk. : alk. paper)
   ISBN-10: 1-4051-2638-8 (pbk. : alk. paper)
   1. Women clothing workers—Economic conditions. 2. Clothing trade—Subcontracting.
3. Clothing workers—Labor unions. 4. Employee rights. I. Hale, Angela. II. Wills, Jane. III. Series.

HD6073.C6T477 2005
331.7′687—dc22

2005006164

A catalogue record for this title is available from the British Library.

The publisher's policy is to use permanent paper from mills that operate a sustainable forestry policy, and which has been manufactured from pulp processed using acid-free and elementary chlorine-free practices. Furthermore, the publisher ensures that the text paper and cover board used have met acceptable environmental accreditation standards.

For further information on
Blackwell Publishing, visit our website:
www.blackwellpublishing.com

Contents

Figures

Tables

Boxes

About the Authors

Maggie Burns has worked on a freelance basis for the past five years undertaking research and evaluation, facilitating North-South advocacy meetings and co-ordinating international campaigns with a Southern base. She is a Director of Women Working Worldwide (WWW) and represents WWW within the Ethical Trading Initiative (ETI) in the UK. Currently she is the NGO co-ordinator for ETI and is working with OXFAM International in the South Asia region to give support to a campaign on the implications of the phasing-out of the Multi-Fibre Arrangement in 2004. Her publications include a report on UK companies operating in Indonesia (CIIR 1999) and 'Effective monitoring of corporate codes of conduct' (CIIR & NEF, 1997).

Angela Hale is full-time Director of Women Working Worldwide, which is based at Manchester Metropolitan University, where she previously lectured in sociology. Angela has also worked for several development agencies, notably Oxfam, War on Want and Womankind. She has published a number of articles on strategies for defending the rights of women workers in a globalised economy, which have built on collaborative work with organisations in Asia and Africa. These include 'Trade liberalisation in the garment industry: Who is really benefitting?' (2002); 'Women workers and the promise of the ethical trade in the globalised garment industry' (with Linda Shaw 2001); 'The Emperor's new clothes: What codes mean for workers in the garment industry' (with Linda Shaw 2002); 'Beyond the barriers: New forms of labour internationalism' (2004); 'Globalised production and networks of resistance: Women Working Worldwide and new alliances for the dignity of labour' (2004); and 'Humanising the cut flower chain: Confronting the realities of flower production for workers in Kenya' (with Maggie Opondo 2005).

Rohini Hensman has worked with trade unions and women's groups in Bombay since 1980, and is a member of the Union Research Group. She is also an active member of Women Working Worldwide and has worked in that capacity on research, consultation and education programmes with women workers in Bombay and Sri Lanka on the social clause, codes of conduct and subcontracting chains. Her publications on these issues include: 'How to support the rights of workers in the context of trade liberalisation' in *Trade Myths and Gender Reality* edited by A Hale (1998); and 'World trade and workers' rights: In search of an internationalist position' in *Place, Space and the New Labour Internationalisms* edited by P Waterman and J Wills (2001). She has also co-authored *My Life is One Long Struggle: Women, Work, Organisation and Struggle* (1984) and *Beyond Multinationalism: Management Policy and Bargaining Relationships in International Companies* (1990).

Jennifer Hurley worked as Research Co-ordinator with Women Working Worldwide on the project exploring garment industry supply chains featured in this book. She has done research into the international garment industry, supply chains and workers' rights for six years. Her interests include the rights of women workers, the global garment industry and research methodology. She has a PhD in International Political Economy.

Doug Miller is Senior Lecturer in Industrial Relations at the University of Northumbria. Since 2000 he has been seconded to the International Textile Garment and Leather Workers' Federation as the Targeting Multinationals Project Co-ordinator, an initiative concerned with the development of global trade union networks within multinational companies, the provision of assistance for national organising drives and campaigns, and support for the negotiation of international framework agreements on global employment standards in the sector. Doug has published recently on European Works Councils and international framework agreements in the textile, clothing and footwear sector.

Lynda Yanz and Bob Jeffcott are founding members of the Maquila Solidarity Network (MSN), a Canadian network that works closely with WWW. MSN promotes solidarity with women's and labour rights groups in Mexico, Central America and Asia working with export processing zone workers to improve working conditions and wages.

Lynda is president of the tri-national Coalition for Justice in the Maqui-ladoras. Lynda and Bob are co-authors of numerous articles on the globalised garment industry and on the strengths and weaknesses of codes of conduct as tools for improving working conditions. Recent publications include: *A Needle in a Haystack: Tracing Canadian Garment Connections to Mexico and Central America* (Toronto: MSN 2000); *A Canadian Success Story: Gildan Activewear: T-Shirts, Free Trade and Worker Rights* (Toronto: MSN 2003); and *Tehuacan: Blue Jeans, Blue Waters and Worker Rights* (Toronto: MSN 2003).

Camille Warren has a Masters in Human Rights and currently works for Women Working Worldwide as a research and outreach worker. She has written articles on the implications of subcontracting for UK workers, on the use of patents in agriculture and on peace issues. She has also contributed to campaigning and awareness raising events in support of the rights of garment workers and workers producing fresh produce for UK supermarkets.

Jane Wills is a Reader in Geography at Queen Mary, University of London and a board member of Women Working Worldwide. Jane's recent research has included enquiry into the use of International Framework Agreements to secure improved labour standards and into community unionism as a means to widen labour organisation to contingent labour markets in the UK. Current research is focused on migrant labour in low-paid employment and the work of the living wage campaign in London. Previous publications include *Union Retreat and the Regions* (with Ron Martin and Peter Sunley, 1996); *Geographies of Economies* (edited with Roger Lee, 1997); *Dissident Geographies: An Introduction to Radical Ideas and Practice* (with Alison Blunt, 2000); and *Place, Space and the New Labour Internationalisms* (edited with Peter Waterman, 2001).

Women Working Worldwide is a small NGO that supports the rights of women workers in industries supplying the world market with consumer goods such as clothing, footwear and fresh produce. Collaborative projects are developed with an international network of trade unions and women workers' organisations, with the aim of increasing the ability of women workers to organise and claim their rights. The outcomes from these projects are used to inform public campaigning and advocacy work in Europe about the impact of the world economy on women workers and the appropriateness of different strategies for defending workers' rights in international supply chains.

Acknowledgements

This book would not have happened without the collaborative work of many people. First of all the contributors themselves, who worked with us as a team to present a comprehensive picture of the garment industry from the perspective of workers. Thank you for all your hard work and also for keeping to deadlines in spite of the weight of other commitments. Also many thanks to all the people in Asia, Bulgaria and Mexico who contributed to the research that underlies much of this book. In particular, we would like to thank project co-ordinators and researchers: Rokeya Baby and Pratima Paul-Majumdar in Bangladesh, Verka Vassileva and Ivan Tishev in Bulgaria, Monina Wong and Jennifer Chuck in Hong Kong, Chanda Korgaokar and Rohini Hensman in India, Aima Mahmood and Nabila Gulzar in Pakistan, Diane Reyes, Marlen Bartes, Gerry Doco-Cacho and Cristi Facsoy-Torafing in the Philippines, Kelly Dent and all members of the Women's Centre and Da Bindu in Sri Lanka and Manee Khupakdee and Jaded Chouwilai in Thailand.

We would like to give a special thanks to Eva Neitzert and Jeremy Anderson who have done a fantastic job in rescuing us from all the work of formatting, checking and amending the final draft of the book. Thanks also go to Steve Kelly of Manchester Metropolitan University Design Department for his contribution to design and artwork and to Edward Oliver who has done a superb job in drawing most of the maps, diagrams and figures.

We would also like to thank the funding agencies that made this book possible, the Community Fund and Department of International Development which supported the action research and the European Community which funded a project that enabled us to spend time publishing the findings. Also thanks go to Francois Beaujolin and the Fondation des

Droits de L'Homme au Travail for valuable contributions to both the work in Asia and the publication itself.

We are delighted that the book is being published as part of the *Antipode* book series and are very grateful to Noel Castree and Jamie Peck for agreeing to support the work. We also owe a big thanks to all the staff at Blackwell, and particularly Jacqueline Scott and Angela Cohen for their work in getting the book into production.

Jane would like to thank Anibel Ferus-Comelo, Jane Holgate, Lina Jamoul, Jeremy Anderson, Paula Hamilton and Claire Frew for all the valuable discussions we have had about the condition of workers and the challenges facing organisers today. Teaching the postgraduate masters degree course at Queen Mary entitled *Globalisation and Development* has also helped in thinking through many of the issues raised in this book. In addition, working with a team of wonderful colleagues has fed into this book in many different ways and I would like to give special mention to Alison Bunt, Kavita Datta, Roger Lee, Cathy McIlwaine, David Pinder, Adrian Smith, David Smith and to Stuart Howard from the International Transport Workers' Federation. Jim Chapman, Agnes and Eric have all fallen over piles of paper that were earlier drafts of this book as they walked and/or crawled round the house, and they deserve a big thank-you for being so supportive both to me and the project.

Angela would like to thank all the management committee and staff of WWW, whose work and inspiration underlie much of the material in this book. Thank you to committee members Diane Elson, Linda Shaw, Peta Turvey, Alana Dave, Barbara Evers, Gerry Reardon, Yvonne Rivers and Helen O'Connell and, of course, to Jane Wills, Rohini Hensman and Maggie Burns who have contributed directly to the contents. Thanks to present and past staff Jess Mock, Joanne Smith and Mary Sayer and above all to Jennifer Hurley and Camille Warren who kept working patiently in spite of the pressure of other demands. I would also like to thank staff of the International Textile, Garment and Leather Workers Federation, the Clean Clothes Campaign and Labour Behind the Label for valuable collaboration and also the staff from Gap who responded positively to our findings. Thanks also go to the Sociology Department of Manchester Metropolitan University, where WWW is based, and in particular to Bernard Leach, Paul Kennedy and Susie Jacobs, for the encouragement given to all WWW's work. Finally thanks go to my family, Richard, Amy and Sonya for constant patience and support, and we would like to dedicate the publication to my grandson Freddy who was born at the same time as this book.

Acronyms

AFL-CIO	American Federation of Labor-Congress of Industrial Organizations
AGOA	African Growth and Opportunity Act
AMT	Ayuntamiento Municipal de Tehuacan
ATC	Agreement on Textiles and Clothing
B2B	business to business
B2C	business to customer
CAFTA	US/Central American Free Trade Agreement
CAW	Committee for Asian Women
CAWN	Central America Women's Network
CCC	Clean Clothes Campaign
CCEIA	Carnegie Council on Ethics and International Affairs
CMT	cut, make and trim
COVERCO	Commission for the Verification of Codes of Conduct
CROM	Regional Confederation of Mexican Workers
CSR	corporate social responsibility
CTM	Confederation of Mexican Workers
EMDA	East Midlands Development Agency
EPZ	export processing zone
ETI	Ethical Trading Initiative
FOB	free on board
FDI	foreign direct investment
FLA	Fair Labor Association
FNV	Dutch Trade Union Federation
FOW	Friends of Women
FROC-CROC	Revolutionary Confederation of Workers and Campesinos

FTZ	free trade zone
FTZGSEU	Free Trade Zone and General Services Employees' Union
ICFTU	International Confederation of Free Trade Unions
ILO	International Labour Organisation
IMSS	Social Security Programme of the Mexican Government
ITGLWF	International Textile Garment and Leather Workers' Federation
JLCT	Junta Local de Conciliación de Tehuacan
KEWWO	Kenyan Women Workers' Organisation
KFAT	National Union of Knitwear, Footwear and Apparel Trades
KWWAU	Korean Women Workers Associations United
LBL	Labour Behind the Label
MEC	Maria Elena Cuadra Women's Movement
MFA	Multi-Fibre Arrangement
MNC	multinational corporation
MSN	Maquila Solidarity Network
NAFTA	North America Free Trade Agreement
NGH	National Group on Homeworkers
NGO	non-governmental organisation
NICWJ	National Interfaith Committee for Workers Justice
NMW	National Minimum Wage
NWDA	North West Development Agency
OECD	Organisation for Economic Co-operation and Development
OI	Oxfam International
OPT	outward processing trade
RFP	request for price
RFQ	request for quote
RMALC	Mexican Action Network on Free Trade
SCMD	State Centre for Municipal Development
SEWA	Self-Employed Women's Association
SITEMEX	Independent Union of Mex Mode Workers
SUTIC	Garment Industry Workers' Union
TAG-MEX	Tarrant Apparel Group
TCSG	Textile and Clothing Strategy Group
TELCO	The East London Communities Organisation
T&G	Transport and General Workers' Union
TIE	Transnational Information Exchange
TNC	transnational corporation
UNCTAD	United Nations Conference on Trade and Development

UNDP	United Nations Development Programme
URG	Union Research Group
USAS	United Students Against Sweatshops
WIEGO	Women in Informal Employment: Globalizing and Organizing
WIN	Women's International Network
WTO	World Trade Organisation
WWO	Working Women's Organisation
WWRE	World Wide Retail Exchange
WWW	Women Working Worldwide

1

Threads of Labour in the Global Garment Industry

Jane Wills with Angela Hale

Introduction

An estimated 40 million workers, most of them women, are employed in the global garment industry. The industry is worth at least US$350 billion (£190 billion) and is expanding year by year (de Jonquières 2004). Clothing production is a major source of employment in many poor countries in the South, and, as such, the industry *could* play an important role in social and economic development on a very large scale. For it to do so, however, there needs to be a massive reconfiguration of the distribution of wealth and power in the industry. Contemporary trends in the organisation of production, reinforced by the re-regulation of the global economy, have made it very difficult for workers to organise and/or to improve their conditions of work.

The women workers upon whom the garment industry depends for its wealth are largely invisible, increasingly distanced from the major brand-name retailers in the industry by complicated chains of subcontracted production. The industry generates immense wealth for those at the top of the corporate hierarchy, while many millions of women are forced to make our clothes in poor conditions, with low pay, forced overtime and insecure hours of work. At present, they are scarcely able to organise at their own workplaces, let alone find the power and resources to try to reshape the evolution of the industry at an international scale. In spite of this, there has been widespread resistance by women workers, supported by a growing network of organisations that have developed to educate and empower women workers in the industry and to link their struggles to wider campaigns.

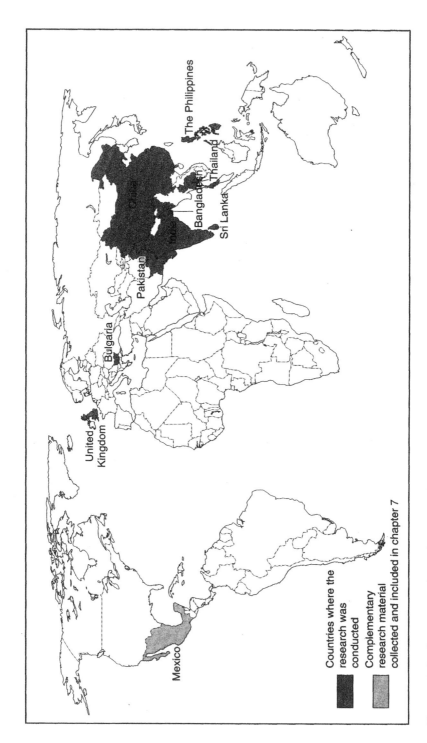

Figure 1.1 Countries where the research was conducted

Women Working Worldwide (WWW) is a small non-governmental organisation (NGO), based in the UK, that has built up a network with some of these women workers' organisations and used these links to inform public campaigning and advocacy work in Europe. WWW supports the rights of women workers in industries supplying the world market with consumer goods such as clothing and footwear, and our particular concern is with the way in which changes in the global economy can have a negative impact on the working lives of women. The aim is to increase awareness of these changes and to support the development of appropriate strategies for defending workers' rights. WWW has been active since 1982 and since then a strong working relationship has been established with workers' organisations throughout Asia, as well as in Africa and Central America (for more information see Chapter 3, this volume). This has enabled the development of collaborative projects on issues of mutual concern. The focus has been on the garment industry, as the most globalised industry employing a majority of women workers.

This book reports on an action research project, co-ordinated by WWW, linking ten local organisations in nine different countries. The research was intended to shed light on the structure of the global garment industry and the scope for resistance in Bangladesh, Bulgaria, China, India, Pakistan, the Philippines, Sri Lanka, Thailand and the United Kingdom. Each local organisation devised research to meet its own needs, while simultaneously contributing to an international collaboration aimed at better understanding the operation of subcontracted supply chains, their impact on employment and their implications for organising. By sharing experiences across national boundaries, local research projects have coalesced to allow a network of workers' organisations to trace the threads of women's labour in the global economy while also supporting ongoing organising activities at local, national and international scales. In addition, the book includes complementary research conducted by similar workers' rights organisations in Mexico. As such, the book provides an insight into the operation of the global garment industry, from the 'bottom up', in ten different countries across Asia, Europe and Central America (see Figure 1.1).

In this book, the findings from this action research are situated within existing knowledge about current trends in the garment industry and it will be seen that the work corroborates analyses provided by more academically based research. At the same time, the research documented here expands on that knowledge by providing information about more hidden operations at the bottom of supply chains, which became visible through the development of a relationship of trust between researchers and workers. In this book, 'bottom up' understandings of international subcontracting chains are also viewed within the framework of changing international trade regulations associated with the phase-out of the Multi-Fibre Arrangement (MFA), which is set to radically change the geographical spread of the industry. The focus of the last part of the book is on the impact of these changes for workers and on the ways in which they and the organisations that support them are challenging the likely negative impact on working conditions. It is argued that the strength and global reach of workers' support networks is such that there are real opportunities for changing the ways in which the industry operates.

By way of an introduction to the book, this chapter provides a brief overview of the development of networked capitalism at the global scale, arguing that networked activism is a necessary response. Over the past decade or so, companies have responded to such activism with the development of new initiatives in corporate social responsibility (CSR). These are outlined, in brief, in the penultimate section of this chapter before we go on to highlight their limitations. In particular, we argue that CSR does not adequately address the impact of subcontracted production and the way that pressure is forced down the chain eroding pay, security and working conditions.

In sum, *Threads of Labour* argues for a renewed focus on the politics and practices of international subcontracting and its impact on workers in garment production and beyond. The book makes the case for tackling the structure of the industry and the way in which subcontracting is managed, rather than focusing on the particularities of production and working conditions in particular parts of the world. Moreover, in the context of the MFA phase out, we argue that there are opportunities to intervene in the evolution of the industry in order to improve the conditions of workers. While we acknowledge that workers are already differentially affected by their position within the hierarchies of international subcontracting chains, and that their power to act is shaped accordingly, we also seek to take an industry-wide approach and consider ways to improve the situation of all those working in apparel production.

Networked Capitalism

Since the 1970s, technological, political and economic developments have conspired to propel a powerful new form of capitalism into view.

Characterised by its networked form and global reach, this model has profound implications for labour. In particular, a growing number of multinational corporations (MNCs) have reconfigured their operations, shedding their in-house production capacity and using subcontracted supply chains to source goods and get them to market. Rather than having their own factories, these companies contract goods and services from their suppliers, retaining only the design, marketing and brand-development functions in-house. Through the management of complex supply chains, leading corporations are able to use geographical differentiation in production costs, legal regulations, trade quotas and labour supply to maximum effect, sourcing their products at the requisite quality and the cheapest price from the most suitable suppliers. Leading MNCs have thus been 'hollowed out' and no longer have to bear the risk of employing large numbers of staff in production. Simultaneously, they have been fuelling competition between subcontractors to keep their own costs low.

In many ways the garment industry is the exemplar of these contemporary trends in global production, and a number of the key texts that have alerted both academics and activists to the implications of globalisation and subcontracted capitalism for the reconfiguration of the world's working class have focused on the clothing industry (see Dicken 2003; Fröbel et al 1980; Gereffi 1994, 1999; Klein 2000; Ross 1997). Low start-up entry costs, labour intensity and the ease of subcontracting all make the garment industry particularly vulnerable to horizontal internationalisation and vertical subcontracting. Although very similar developments have taken place in the electronics industry, auto manufacturing, toy production and the horticultural sector, in many ways, the garment industry has pioneered the new trends (Barrientos et al 2003; Cook et al 2004; Dicken 2003; Harvey et al 2002; Holmes 2004; Raworth 2004).

A commodity or supply chain approach has become dominant in academic accounts and analysis of these patterns of global manufacturing. Gereffi (1994, 1999) has argued that the global apparel industry is now characterised by buyer-driven commodity chains, as distinct from more traditional producer-driven chains. By this he means that garment industry supply chains, which involve a wide range of component inputs and the assembly and distribution of finished goods, are determined by the retailers and brand-name merchandisers. He argues that these buyers create the geo-economy of garment industry chains by having the ability to select suppliers in different parts of the world. In addition, the governance of such chains is characterised by the power of those at the top. Gereffi (1994:97) uses the term governance to refer to the 'authority and power relationships of the chain.' Subcontracting in buyer-driven chains like those in the garment industry concentrates power with the buyers rather than with those making the goods.

This analytical framework has been used widely, by a range of academics and activists, to make sense of the changing geography of global production. As might be expected, scholars have gone on to add to the model in order to incorporate additional factors and emphases. Indeed, Gereffi and his various co-authors have themselves gone on to explore the ways in which value is produced and captured within commodity chains and the implications of this for industrial policy and efforts to upgrade production capacity in different parts of the world (Gereffi 2003; see also Kaplinsky 2000). In recent years, a number of geographers have used the model to engage with debates about globalised production but have sought to avoid the implied linearity of the focus on chains, replacing it with attention to networks. In his research on global production, for example, Dicken (2004:15; see also Dicken et al 2001) has used the notion of global production networks that are 'a nexus of interconnected functions and operations through which goods and services are produced, distributed and consumed.'

This approach has the advantage of widening the net to include all those who are involved in the processes of commodity production, distribution and consumption. Moreover, it also enables a widening of the analysis of power relations to embrace institutional, labour and civil society actors and their networks, and their potential impact on the nature and location of production, distribution and exchange. As such, a focus on networks can also help to identity points for intervention and political action in particular corporations and industrial sectors. Taking up this agenda, Smith et al (2002:47) argue that labour has been strangely neglected. Workers appear only 'as passive victims as capital seeks cheap labour' in much of the commodity chains literature. Moreover, they go on to suggest that it is important to pay particular attention to labour processes and political organisation in the constitution of commodity chains:

> We would contend that labour process dynamics strongly influence wealth creation and work conditions within any one node and across a chain. In addition, we would argue that organised labour can have an important influence upon locational decisions within and between countries thereby determining in part the geography of activities within a value chain (Smith et al 2002:47).

It is this call for attention to labour that we take up in the rest of this book. Not only has labour been neglected in the analysis of commodity chains, but little work has been done to explore what these new forms of capitalist production mean for labour politics and practice.

Labour movement responses

The development of internationally networked, subcontracted capitalism has had a devastating impact on traditional trade union organisation. Successful collective bargaining requires that there are two parties to bargain: one with (potential) collective strength and the other with the means to concede change (or otherwise). When a company owns a factory there is a direct relationship involved in this negotiation of power: workers and employers need each other, and have to co-operate with each other to some extent at least. But in subcontracted capitalism, those with real power over the contracting process—the ultimate employers of all those involved in any particular supply chain—are generally hidden from workers and located many thousands of miles away overseas. Managers of these supply chains are not directly responsible for the workers and are often less than fully dependent on them for the production of goods. This limits the scope for collective bargaining over the terms and conditions of employment. If workers were to demand improvements that put up costs, it is likely that they would end up losing their jobs, as the contract would be shifted elsewhere.

Indeed, even in cases of workers' protest that have involved international solidarity action, workplace organising has often resulted in the leading brands and retailers reconfiguring their supply chains, to source their goods from elsewhere (see Bonacich 2000; Bronfenbrenner 2000; Cravey 2004; Traub-Werner and Cravey 2002). This model of capitalism increases competitive pressures on suppliers, and even the largest manufacturers are under severe pressure to keep costs as low as they can. While there might be scope for workers in these larger factories to win small improvements in the terms and conditions of work, dramatic improvements will depend on those at the top of the chain.

In this context, the prevailing model of workers' organisation that is focused on creating workplace trade unions needs to be overhauled. Traditional trade unionism makes less sense than it did in the past, not least because organising at the workplace is no longer enough. Manufacturing workers in particular, need to be able to challenge the impact of subcontracting that is controlled beyond their own workplace.

If workers are to bargain to improve their conditions of work, the competitive contracting environment and the unequal power relations on which it rests need to be tackled. This involves understanding where power lies in the subcontracting chain and in particular, how the control exercised by buying companies over prices and production schedules limits the ability of suppliers to respond to workers demands for improvements in pay and conditions. Understanding this geographical differentiation of power in subcontracting chains is critically important if we are to develop new ways to improve workers' conditions of life.

The geographical distance involved in networked capitalism means that workers are not only isolated from their ultimate employers but also from the consumers of their products. Northern consumers never see the workers who make their clothes and rarely make any connection between the prices they pay for fashion items and the quality of life of workers. Furthermore, the producers and consumers involved do not share any political institutions that could be used to put pressure on those at the top of the contracting chain (for an argument that they should, see Monbiot 2003). This contrasts with the public services where services are typically contracted, provided and consumed in a shared geographic location. In the public services sector in North America, subcontracting has similarly been used to keep costs down by reducing pay, eroding working conditions and reducing the quality of service provision. But here living wage campaigns comprising coalitions of community-based groups and trade union organisations have highlighted the impact of subcontracting on workers, and made political demands for the state to regulate, and for employers to act, in order to improve the terms and conditions of those doing the work (see Pollin and Luce 1998; Reynolds 2001; Walsh 2000; Wills 2004). Such campaigns can construct a community of interest between workers, community-based organisations and service users regarding the benefits of higher wages and better conditions, and then deploy their collective political power against the local state and employers to this effect.

Subcontracting thus poses sectorally differentiated political challenges that depend, at least in part, upon geography. Geographical distance needs somehow to be overcome so that a community of interest and solidarity can be constructed that is able to recast the way in which subcontracting in manufacturing takes place. In order to do this, activists will need to develop a sophisticated understanding of the networks of capital involved, where they are grounded, and where political action can have an effect. As Massey (2004:11) puts it:

Different places are of course constructed as various kinds of nodes within globalisation; they each have distinct positions within the wider power-geometry of the global. In consequence, both the possibilities for intervention in (the degree of purchase upon), and the nature of the potential political relationships to (including the degree and nature of responsibility for) these wider constitutive relations will also vary.

Thus, to organise successfully in the global economy, workers in buyer-driven supply chains need to be part of a new kind of political organisation or set of networks with the necessary political tools to change the way in which capitalism works. Such political organisation requires awareness of geographically differentiated economic and social relations, and the development of transnational links between workers in producing locations and consumers and activists in the key markets and home ground of the main MNCs. This kind of networking activity has developed over the past two decades and has gradually had greater leverage over conditions in the garment industry. Women Working Worldwide sees itself as part of this activity, having developed relationships with emerging organisations of women workers in the industrialising areas of the South since the early 1980s and then sought to build links between them and trade unions and consumer-based organisations in the main markets of Europe. These organisations include the Clean Clothes Campaign (Europe), the Maquila Solidarity Campaign (Canada), and United Students Against Sweatshops (US), all of which have focused their campaigns around labour conditions in the garment industry and have sought in different ways to forge greater transnational solidarity between workers and consumers (for more information, see Chapter 3, this volume, and Johns and Vural 2000).

These and other similar organisations are also engaging in debates about the need for corporations to act responsibly, about the injustices of world trade and the battle against neo-liberal models of political economy. As Heintz (2004:225) argues in his review of the challenges facing the anti-sweatshop activists in North America, it is necessary to: 'contest the current structures of the global economy in ways that expand opportunities and protections for the most vulnerable segments of the world's labor force.' This is the agenda not only of the emerging workers' rights networks but also of the global justice movement more generally, which recognises the plight of workers in international supply chains as one of the manifestations of the devastating impact of current models of globalisation (see Fisher and Ponniah 2003; Monbiot 2003; Wainwright 2003).

Activist organisations have drawn attention to the situation of the women workers upon whom the garment industry depends for its wealth. However, although a body of work on gender and industrialisation has highlighted the plight of women workers in globalised production (Elson and Pearson 1998; Enloe 1990; Perrons 2004; Standing 1989), workers have been largely invisible in academic and policy debates associated with the analysis of commodity chains (for important exceptions, see Barrientos et al 2003; and the research completed by Women in Informal Employment Globalizing and Organizing (WIEGO), some of which appears in Lund and Nicholson 2004). Moreover, in both academic and activist circles, there is a danger of representing women as exploited and disempowered workers in sweatshop economies with no voice of their own (for more on this critique see Kabeer 2000, 2004). This is despite the fact that new organisations—often located outside the workplace—have developed in most areas of export-oriented production in order to support, educate and empower women workers in the industry. In many cases, these organisations have developed, at least in part, as a result of the difficulties of organising as unions. Where state repression, employer hostility and/or union weakness have made it difficult or impossible to organise in the workplace, such organisations have come to play a similar and/or complementary role (Rowbotham and Mitter 1994). It is critically important that ways are found for activists involved in international campaigns to link effectively with these new organisations and trade unions so that they can work together with workers in voicing demands.

As an example of new forms of women workers' organisation and the context in which such organising takes place, it is valuable to look at a country like Bangladesh, where garment production has transformed the landscape of employment and labour organisation. The industry has grown rapidly since the 1980s and there are now some 3280 factories and some 1.8 million workers (almost all of them women) employed in the sector (Kabeer 2004:15). It is estimated that less than 10% of these workers are engaged in the largest factories in the country's two export processing zones (EPZs) where conditions are most favourable to workers. Most are employed in factories and workshops with less than 500 workers, or are working at home. While some of the workers in these smaller factories and workshops, and some homeworkers, will be sewing clothes for export, they are likely to be engaged in sub-subcontracted operations, with little knowledge of, or relationship to, those at the top of their production network or chain. In the particular context of subcontracted garment production—much of it informalised—trade unionists in Bangladesh have struggled to establish workplace organisation. But this does not mean that the workers involved are not challenging many aspects of industrial life. Indeed, the organisations that have been developed, such as Karmojibi Nari (a partner in the WWW research work reported in this book; see Chapter 4, this volume), are supporting women workers and working alongside existing trade union organisations. Organisations like Karmojibi Nari reflect the experiences of women workers, highlighting their need for particular forms of workplace and community organisation, as Kabeer (2004:23) explains:

> Such organisations bring with them the recognition that women workers' exercise of agency in the workplace is unlikely to take the form of the heroic mass struggles that make up trade union lore. Engaged in unceasing, individual struggles on a daily basis to combine their domestic chores with waged labour, and to negotiate their way in a world hostile to the idea of women working for pay, their agency in the workplace takes a lower key and less confrontational form . . . mass collective action through the trade union movement remains a remote possibility.

Likewise, in Central America, it is estimated that trade unions represent less than 1% of *maquila* workers outside Honduras, and it is women workers' organisations that have begun to fill the gap in labour representation (Prieto and Quinteros 2004). Successful initiatives here, as in other parts of the world, have comprised: the establishment of centres to serve those in the key areas of export production; the establishment by faith organisations of trusted support groups for workers; and the provision of routes to connect with workers by social welfare initiatives offering housing, medical, educational and legal advice. As Shaw (2002:54; see also Dannecker 2000; Hale 2004) explains: 'Some of the most successful organising work has been done by women focusing on community issues and issues that connect the community and the workplace—housing, childcare, transportation, safe drinking water, health, environmental protection.' Once such community groups are established, they have shown themselves able to build bridges to any existing trade union organisations and/or to promote new ways of organising that are sympathetic to the circumstances of many women workers.

In this regard, it is significant that such methods are now being developed to organise workers in the North as well as the South. Where workers have not had the opportunity to establish traditional trade union organisations, or where they face too many barriers, new collective organisations are being developed. In North America, a number of workers' justice centres have been set up to reach, support and mobilise groups of low-paid workers, many of them migrants, who have weak associations with their workplaces and stronger affiliations outside (Gordon 2001; Ness 1998). Likewise, new coalitions of trade unions and community-based organisations are reaching and organising workers beyond the walls of the workplace. This is most evident in the considerable number of community-union living-wage coalitions in the US (see Harvey 2000; Walsh 2000; Wills 2001a, 2004). As the landscape of industrial employment has changed, so new organisations have been created, and this book documents some of the efforts that are being made by a number of these groups as they work, often alongside trade unions, to tackle the power imbalances created by subcontracted production.

Corporate Social Responsibility

It was during the 1980s, when deindustrialisation accelerated in the North, that corporate experiments in subcontracted production really took off. Fuelled by the dominance of neo-liberal conservatives at the helm of the global economy, companies argued that globalisation was good for business, even if it inevitably involved a 'race to the bottom' in labour, social and environmental standards. Indeed, following the principles of laissez-faire economics, the gurus of the new capitalism argued that the favourable investment conditions afforded by cheap labour were an advantage to the developing world, facilitating ever greater development and 'trickle-down' to the poor. By subcontracting production, corporations realised that they could secure greater flexibility and profitability without owning productive capacity as had been necessary in the past.

It was not really until the 1990s that coordinated opposition to this model developed on an international stage. In the wake of a rising tide of protest, witnessed on the streets of Seattle, Genoa and Cancun, social movements began to coalesce in their battle against neo-liberal capitalism. Huge demonstrations, corporate campaigns, the development of social forums and fair-trade initiatives have helped to shift the debate and put the corporations under some pressure to change. In this context, social and environmental responsibility have become 'bottom-line issues', critical to the reputation of a company, the value of its brands, its attraction and retention of staff, and its success in the market place. Political action has thus created some space in which to manoeuvre against corporations in the interests of labour, community and environmental protection (Adams 2002; Rock 2003).

Faced with organised opposition, a number of companies have been prepared to publicly recognise their responsibilities to those they employ and have developed standards of corporate social responsibility (CSR), to which they and their suppliers are supposed to adhere. Many have adopted corporate codes of conduct that set out minimum standards for conduct along their supply chains (see Diller 1999; Hale 2000a; Hughes 2001; Jenkins et al 2002). In addition, a very small number have negotiated framework agreements with the global union federations, allowing for a more rigorous monitoring and implementation regime (see Miller and Grinter 2003; Wills 2003a). Organisations like the multi-stakeholder Ethical Trading Initiative (ETI) in the UK and the Fair Labor Association (FLA) in the US have been valuable in pooling resources and expertise in the implementation, monitoring and verification procedures of codes (see Chapter 3, this volume).

On the face of it then, alliances of political campaigners, trade unionists and consumers appear to have won considerable concessions, bringing at least some corporations to social account. Moreover, by encouraging corporations to develop the policies and practices of CSR, campaigners have strengthened their armoury to defend workers' rights, not least because corporate hypocrisy makes much better news than straightforward bad practice. At the same time, however, the plethora of CSR initiatives has done nothing to alter the facets of the global economy that cause the problem of poor and declining standards of employment in the first place. Intense competition, particularly in highly price-sensitive markets such as those for garments, food and electronics, has not gone away. Nor indeed, has the need to deliver ever greater shareholder value. In pursuit of both, large corporations designing and selling manufactured goods will inevitably look to reduce costs, and subcontracting production is a very effective route to that end.

As the research in this book illustrates, corporate supply chains in the garment industry now involve multiple layers of subcontracted production, some of it visible but much of it hidden. Even if a company has a code of conduct or an exemplary policy to promote CSR, it is extremely difficult to know exactly what is happening on the ground. Workers will often be unaware of who they are working for, and are even less likely to know about any codes of conduct or their implementation (and for details of previous research and activism by WWW on codes of conduct, see Hale 2000a; Chapter 3, this volume). Thus while WWW has wel-comed the development of CSR, corporate codes of conduct, ethical trading initiatives and the opportunity to take part in new attempts to improve labour standards, we remain very cautious about the impact such measures will have. This book highlights the systemic causes of poor working conditions and suggests that subcontracted chains of production and the wider political-economic environment that this kind of production helps to create, means that *most* garment workers, and perhaps an increasing proportion of them, are deprived of good working conditions and rights in employment.

Threads of Labour

The contributors to this volume have all been involved in the work of WWW, as staff or management or as collaborators on particular pro-jects. The book is very much a collective product arising from our networking activities over the past 20 years. In what follows, Chapter 2 explores the global operations of the garment industry in more detail, focusing on the changing nature of supply chains and the key processes that impact on those doing the work. Following this overview, Chapter 3 introduces the research that has taken place in the context of WWW's involvement in what is termed a 'new labour internationalism' that has developed since the 1980s, linking women workers' organisations in poor countries in the South with growing consumer activism in the North. This chapter addresses the emergence of new forms of workers' organisations, North–South networks and campaigns for the rights of garment workers, the development of corporate codes of conduct and the need to include workers in their development and implementation. The chapter concludes by explaining why women workers' organisa-tions elected to research subcontracting chains in more detail, highlight-ing the need to understand the structure of the industry in order to change it.

The organisations involved in the research are introduced in Chapter 4. Here the principles of action research are elucidated, before the research and its outcomes in each of the nine countries involved are outlined. It is argued that locally implemented, internationally co-ordinated action research provides a methodological framework for conducting research in an era of globalisation.

Chapters 5, 6 and 7 draw on material collected by the project partners and collaborators in ten different countries. The main trends identified across the projects in Asia and Europe are summarised in Chapter 5. This chapter develops an approach to conceptualising supply chains in the garment industry from the perspective of workers, arguing that much happens 'below the water line,' where it is difficult to trace what goes on. The chapter also highlights the main issues for workers employed in garment production, drawing on examples from the action research. Chapter 6 then explores the particularities of developments in the UK—where apparel is a declining industry—and Chapter 7 looks at recent changes in Mexico, further illustrating the similarities and differences in the experiences of workers in different parts of the world. Chapter 8 goes on to draw out the lessons of the research for the battle to defend and extend workers' rights in the industry. And, crucially, the chapter argues that the strategy best able to meet the needs of workers will depend on their position in the supply chain, which is linked to their employment security and status, the extent to which they are covered by local labour law and their ability to campaign for new national and international regulation. Chapter 9 sets these debates in the context of likely future changes in the industry focusing on the implications of the end of the MFA—and its associated trade regime—in 2005.

In sum, *Threads of Labour* explores the impact of international garment industry supply chains from the 'bottom up.' The book seeks to contribute to debates about the globalisation of the economy, the operation of international commodity chains and new developments in labour organising from the perspective of the workers involved. Drawing on internationally co-ordinated but locally developed action research has allowed us to highlight local experiences alongside global trends. We have sought to embody supply chain analysis, and bring it to life by looking at the experiences and situation of some of the workers involved in the contemporary garment industry. The action research data has already been used by local organisations that support women garment workers, informing educational programmes, political action and organising work. This book seeks to share this experience more widely, highlighting the way in which action research can enhance academic debate by developing new insights at the same time that it is used to change the world from below. By publishing *Threads of Labour* we also hope to extend the political reach of the research by contributing to ongoing debate and action about the need to reconfigure economic power relations to the benefit of workers, their communities and poorer nations in the South. The book aims to contribute to ongoing efforts to recognise and improve the position of women working in global production networks, in and beyond the garment industry.

2

The Changing Face of the Global Garment Industry

Jennifer Hurley with Doug Miller

Introduction

This chapter explores the operation of the global garment industry and current developments in the sector. It sets the scene for the material in much of the rest of the book that draws on WWW research, taking a 'bottom-up' or 'worker's-eye' view of the industry. The major trends identified and explored in this chapter were borne out by the research on the ground, but the research also revealed some new information about the local end of global supply chains, which will be reported in greater detail in Chapter 5. By linking the garment industry at the global level, in this chapter, with the research findings at the local level, in Chapter 5, we open up and explore the complex interrelationship between the more abstract elements of the industry—such as international regulations and company sourcing decisions—and the very concrete impacts that these decisions have on the daily lives of individual women working in the garment industry.

To set the scene at the global level, this chapter first describes the nature of the global textile and garment industry. We use a supply-chain approach that allows us to link business decisions at a global level to the experiences of individual women workers at the local level. Drawing on a case study of the Gap's supply chain, we illustrate the complexities of subcontracting that are explored in greater detail in Chapter 5. This chapter then examines contemporary trends in the garment industry, looking specifically at lean retailing and e-commerce, which are altering its structure and, as the research findings show, intensifying the pressure on workers at all levels within it. Finally, the chapter looks at the way in which the regulation of trade affects the industry and provides another case study to highlight the ways in which such regulation has shaped industry practices and in turn impacted on workers.

Making Sense of the Global Garment Industry

As a relatively low-cost labour-intensive activity, export garment assembly is one of the few industries in which developing countries can offer comparative advantage in manufacturing, particularly through labour costs. For the governments and entrepreneurs of developing countries, the industry has been seen as a development lynchpin, opening doors to foreign investment, bringing in foreign exchange earnings and, ideally, acting as a gateway to more value-added industries and services. Garment industry investment opportunities have been viewed as the first step into the international trading arena and the path to export-led economic growth. Many developing countries have attempted to make full use of the industry's potential and developing countries now account for 70% of world exports of clothing (Diao and Somwaru 2002:129). Although the global garment sector accounts for only 3.2% of world manufacturing exports, the world apparel trade has increased some 128-fold in the last 40 years (Someya, Shunna and Srinvasan 2002). With a current value of US$201 billion (2002), and a prognosis that, on present trends, the world's five major markets (US, EU, China, India and Japan) will more than double in the next decade, it is understandable why many buyers and potential sellers are keen to invest in this business (Flanagan 2003:23; see figure 2.1).

Advocates of globalisation point to the contribution which the industry can make in terms of exports, employment and value added. In Bangladesh, clothing accounts for 75% of the country's total export earnings; in Mauritius the figure is 64%, in Sri Lanka 50%, and in Tunisia 40% (Appelbaum 2003:17). In terms of employment, Bangladesh has 1.6 million workers, almost 65% of its total workforce, engaged in the clothing sector. In Tunisia and Morocco, 40% of the national labour force are employed in textiles and clothing. In Turkey, the figure is 34% (Someya, Shunnar and Srinvasan 2002). The share of apparel in the total added value of merchandise exports is also considerable in certain countries—in Bangladesh the percentage is 55%, in Turkey 28%, in Pakistan and Morocco 20% (Applebaum 2003). On the face of it, such statistics might appear to underpin this orthodox 'development model' of the globalisation process, but they mask the specific structural conditions that determine and 'rig' the global apparel market in favour of the buyers.

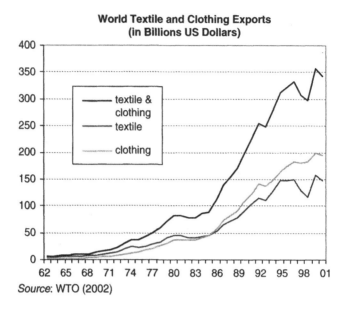

**World Textile and Clothing Exports
(in Billions US Dollars)**

Source: WTO (2002)

Figure 2.1 World textile and clothing exports

The garment industry can be seen as manifesting the classic pattern of 'global shift' in manufacturing as production bases move from one country to another country and from one region to another region: from high-cost to low-cost production locations (Dicken 2003). However, this shift does not operate in a free market. The economic and political forces that structure the global garment industry favour the strongest actors, and this impacts on the global distribution of the industry.

Using a supply chain approach

The past three decades have seen significant changes in the structure and organisation of the industry. At the global level, there has been increased consolidation of power among the biggest players—the retailers, branded manufacturers and marketers—accompanied by the development of more complex networks at the local level (Gereffi 1994). Attempting to conceptualise such a diverse, complex and internationally dispersed industry presents serious theoretical challenges. As outlined in Chapter 1, however, one approach that has shown the flexibility necessary to capture the complexity of developments is the supply chain or commodity chain approach (Gereffi and Korzeniewicz 1994;

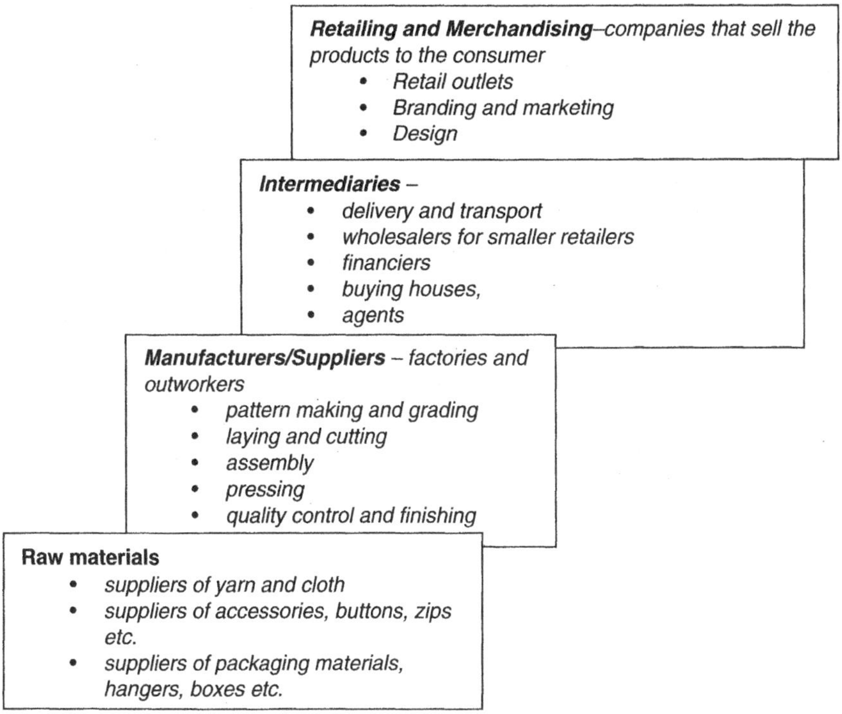

Retailing and Merchandising–*companies that sell the products to the consumer*
  - *Retail outlets*
  - *Branding and marketing*
  - *Design*

Intermediaries –
  - *delivery and transport*
  - *wholesalers for smaller retailers*
  - *financiers*
  - *buying houses,*
  - *agents*

Manufacturers/Suppliers – *factories and outworkers*
  - *pattern making and grading*
  - *laying and cutting*
  - *assembly*
  - *pressing*
  - *quality control and finishing*

Raw materials
  - *suppliers of yarn and cloth*
  - *suppliers of accessories, buttons, zips etc.*
  - *suppliers of packaging materials, hangers, boxes etc.*

Figure 2.2 Simplified functions of a garment supply chain

Sturgeon 2001). This approach provides a framework not only for mapping out the different players in globalised industries, but also for revealing the significance of negotiations between firms and individuals at different stages of the chain (Wood 2001). The garment industry does not operate through anonymous markets but through political and economic relationships at every level from global trade negotiations through to the employment of homeworkers.

At the most basic level, supply chains are made up of all the stages involved in the production and sale of a specific product, from sourcing the raw material to its final destination in a shop. The chain can be broken down into four key functions: raw material supply, intermediary roles, manufacturing and retail. Within each of these functions are different roles and tasks, as illustrated in Figure 2.2.

There is frequently a great deal of separation between the various tasks and actors in any such chain, with each stage being carried out by different companies or individuals. One firm may weave textiles, while another firm organises logistics, and an agent may source trimmings, such as buttons and thread. The strength of the supply chain approach, however, is that it does not just see these players—whether they are individuals, governments or multinationals—as independent, free-floating agents, but as actors who are linked through complex networks and legislative and financial ties, as well as across space.

Though it lacks a well-defined theoretical framework (Gereffi et al 2001:3), the supply chain approach provides useful concepts that enable cross-border networks to be explored from a variety of perspectives. The work of Gereffi is particularly useful for our analysis, because of his focus on the significance of power relationships within the chain. The garment industry is portrayed as a 'buyer-driven' chain and his approach involves looking at how 'lead firms'—like Gap—govern their supply chains and how relationships are organised within such chains (Gereffi 1994, 1999; Humphrey and Schmitz 2001). In buyer-driven chains (as distinct from producer-driven chains) it is the retailer that buys the clothes from the manufacturer that has the power to dictate turnaround times, prices and quality (Gereffi 1994:55). This means that large buying companies are more likely to have greater power in the supply chain. They get this power from their position in the market—literally, how big they are—and how much their marketing activity contributes to the profit they make. For example, jeans sold with Levi's brand name cost more than jeans with a generic brand name, so Levi's gain power from the profit they are able to make from their brand name.

Supply chain analysis has also been adopted by those looking at the industry from a developmental perspective. Here the focus is on the potential for industry upgrading, and, in particular, the extent to which smaller developing-world manufacturers can become more self-sustaining and move further up the chain (Gereffi 1999; Kaplinsky 2000). Work has also been done on understanding the linkages between various points in the chain, the 'drivers' that help generate success and the barriers that block progress in the chain (Dicken and Hassler 2000; Hassler 2000). If the industry can be upgraded, adding more value through production in any one place, this has major implications for wealth creation and further economic and social development.

This approach frames much of the argument in the rest of this book. We can use it to examine the role of multinationals or lead firms in structuring chains, the use of power and the patterns of governance in these chains, the impact of technology, the importance of gender in providing flexibility within chains and the impact of national governments and international regulation.

The changing structure of garment supply chains

In the past decade, there has been a noticeable restructuring in garment supply chains, which has increased the power and profits of lead firms. Mergers and acquisitions among the biggest players have given these companies greater power to shape the industry. Wal-Mart, for example, the world's largest multinational has an annual turnover of nearly $118 billion. Together Wal-Mart and K Mart (turnover p.a. $32 billion) outsell all department stores combined, and their purchasing decisions shape much of the apparel industry (Retail Forward Inc 2003). With the ten largest clothing retailers accounting for nearly two-thirds of all apparel sales in the US, this consolidated buying power vastly increases retailers' ability to put more pressure on the manufacturers in their chains. They have used this power to push down prices and insist on fast turnaround times for delivery to the market. A significant development has also been the rise of private labels owned by the retailer. While retailers typically keep 50% of the price of brand-name garments, they are able to keep 80% of the price of their own private-label products (Sweatshopwatch 2003).

There are three basic types of lead firm in garment industry supply chains: retailers, marketers and branded manufacturers (Gereffi 2001:1625). A glance at the top twenty clothing companies (Table 2.1) reveals that they are all headquartered unsurprisingly, in the world's major clothing markets—the US, EU and Japan. Virtually all are now best described as *merchandisers*. This means they are brand owners that either do not own any production or are in the process of divesting their manufacturing in favour of outsourced offshore production. Benetton, Nike, Adidas, Tommy Hilfiger, Liz Claiborne, Polo Ralph Lauren are classic merchandisers with centralised marketing, design and finance functions at their headquarters. Companies such as Vanity Fair Corporation and Levi Strauss are examples of branded manufacturers that own some manufacturing capacity but are in the process of cutting back on this. As an example, Vanity Fair owns Wrangler and Red Kap and runs factories in Central America, although most of its original manufacturing in the US has been closed down. Levi Strauss, with 501 suppliers worldwide (Fair Labor Association 2003) has embarked on a strategy of closing down its remaining owned facilities in the US, Canada and Europe, and now has just a handful of factories left worldwide (Payne 2002). Triumph International is an example of a multinational which has long maintained its own manufacturing but has increasingly outsourced

Table 2.1 Major clothing companies in the industrialised countries

Ranking	Company	Country of Origin	Product	Turnover in 2001 Million €	Turnover in 2002 Million €	% Change
1	Sara Lee Corp Brand App	USA	Knitwear	8672.0	6826.0	−21.29
2	VF Corporation USA	USA	Jeanswear	6162.4	5376.0	−12.76
3	Jones Apparel Group Inc	USA	Womanswear	4547.90	4590.60	0.94
4	Levi Strauss & Co	USA	Jeanswear	4484.90	4384.60	−2.24
5	LVMH-Gruppe Clothing	France	Prêt-à-Porter	3612.0	4194.0	16.11
6	Zara-Ind Dis. Text.	Spain	Menswear	3249.9	3974.0	22.28
7	Liz Claiborne USA	USA	Clothing	3850.5	3931.40	2.1
8	Fast Retailing	Japan	Clothing	3143.50	2624.0	−16.53
9	Ralph Lauren—Polo	USA	Clothing	2485.20	2499.70	0.58
10	Shimamura	Japan	Womenswear	2228.30	2339.60	4.99
11	Kellwood Co	USA	Clothing	2547.8	2331.60	−8.49
12	Adidas Salomon AG	Germany	Activewear	2212.0	2288.0	3.44
13	Onward Kashiyama Co	Japan	Menswear	1894.20	2231.10	17.79
14	Tommy Hilfiger	USA	Menswear	2095.5	1998.7	−4.62
15	Benetton Clothing	Italy	Knitwear	2097.6	1991.8	−5.04
16	World Apparel	Japan	Womenswear	1574.40	1972.0	25.25
17	Marzotto-Abbigliamento	Italy	Menswear	1410.0	1700.0	20.57
18	Triumph International	Switzerland	Clothing	1655.1	1625.0	−1.82
19	Warnaco Group—Clothing	USA	Underwear	1866.1	1578.8	−15.4
20	Five Fox Group	Japan	Clothing	1541.7	1524.0	−1.15

Source: Euratex Bulletin 2004

production, as well as acting as a contractor to private labels such as Marks and Spencer and C&A and brands such as Esprit and Adidas. Overwhelmingly, the trend is towards the 'new economy' merchandiser business model, whereby focus is placed on the development of brand image through marketing and design, while production, packaging and delivery are left to other companies (Klein 2000).

Big global companies have gradually reduced their manufacturing to refocus their core business on service-related functions (Gereffi 2001:1627). While subcontracting the labour-intensive and competitive activities of production, packaging and transportation, they have also streamlined their businesses in order to focus on the areas of the garment industry that generate the highest profit levels, most notably design, marketing and retail; the so-called 'intangibles.' The growth of private labels is one aspect of this, and the US-based market information company NPD estimates that private label sales now represent 51% of apparel sales in the mass merchandise segment of the retail market (Barrie 2003:20).

As the major retailers/merchandisers no longer have their own manufacturing bases, they are dependent on other manufacturers for their production needs. However, the way in which they source their products is in constant change. In line with the move towards streamlining their business focus, these key global players are now also downsizing the number of manufacturers with whom they do business in an attempt to make their supply chains shorter, as well simplifying and centralising the co-ordination and management of the manufacturing process. As a result, pressure has built up from the global retailers, marketers and branded manufacturers for large multinational manufacturers to provide a 'full-package' service, where the contractor—the manufacturer—co-ordinates all functions of the chain, from sourcing raw materials, project management, delivery and distribution (Bair and Gereffi 2001; Flanagan and Leffman 2001; Gereffi 2001; International Labour Organisation 2000; see also Chapter 7, this volume). The top lead firms globally are developing more long-term partnerships with these transnational manufacturers who are increasingly providing a 'one-stop shop' solution (International Labour Organisation, 2000:88). The design is done by the lead firms, and the orders are then passed on to the manufacturer who is responsible for all aspects of production. As a result, the lead firms appear to have 'flat' supply chains involving relatively few contractors (Gereffi 2001:1627).

However, this 'flattened' supply chain is only the tip of the iceberg. While the relationship between the lead firm and one of their 'lead manufacturers' appears clear and uncluttered, below this are complex supply chains to other garment producers that present a far more complex and intricate picture. The research outlined in Chapter 5 confirmed that consolidation at the top of the pyramid has been accompanied by a lengthening and diversification of the supply chain below the level of the transnational manufacturer. Since these levels of the supply chain are hidden, the structure is best characterised not as a pyramid but an iceberg, a model developed with Stephanie Barrientos during the course of a WWW seminar on both garment and horticulture supply chains (Women Working Worldwide 2004). The dense and complex webs at the bottom end of the chain are invisible not just to outsiders such as government monitors, but also to the retailers that issued the order and sometimes even to the manufacturers that subcontracted the order (see Figure 2.3).

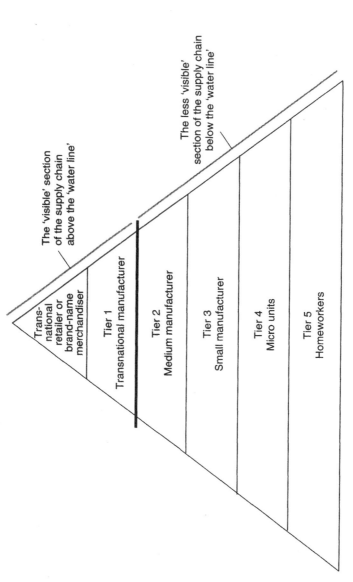

The 'visible' section of the supply chain above the 'water line'

The less 'visible' section of the supply chain below the 'water line'

Trans-national retailer or brand-name merchandiser

Tier 1
Transnational manufacturer

Tier 2
Medium manufacturer

Tier 3
Small manufacturer

Tier 4
Micro units

Tier 5
Homeworkers

Figure 2.3 The pyramid/iceberg model of the supply chain
Source: Women Working Worldwide (2004)

Taking the iceberg model it is possible to look more closely at how the supply chain operates. Above 'sea level,' the global retailers/merchandisers are creating long-standing alliances with a limited number of large multinational manufacturers. These alliances are tightly structured, relatively stable, long-term relationships characterised by a simple structure and clear communication channels. The manufacturers in Tier 1 are multinationals in their own right and have the capacity to provide the extended services required by the retailers and brand marketers. They supply many international clients, have production bases in a number of different locations, and some have a vertically integrated supply chain. As an example, Nien Hsing is a Taiwanese-owned denim and jeanswear manufacturer that supplies major brands and retailers such as Lee, Levi Strauss, K-Mart, J C Penney, and Bugle Boy, from factories in Nicaragua, Mexico, Taiwan, Lesotho and Swaziland. Similarly, Ramatex Berhad, a Malaysian-owned multinational manufacturer, supplies knitwear to major customers in Europe and the USA, such as Nike Puma, Adidas, Otto Versand, Target, Wal-Mart and Sears Woolworth (Mollet 2001). Likewise, Ramatex is an example of a multinational which runs a vertically integrated operation producing its own yarn and knitted fabric and assembles garments in China, Malaysia, Namibia, Brunei, Cambodia and Mauritius. This company provides a full-package service for multinational buyers.

The complexity of garment sourcing at Tier 1 is illustrated by the following quote from the chairman of Li and Fung, a multinational trading house (Magretta 1998:108):

> Say we get an order from a European retailer to produce 10,000 garments. For this customer, we might decide to buy yarn from a Korean producer but have it woven and dyed in Taiwan. So we pick the yarn and ship it to Taiwan. The Japanese have the best zippers and buttons but they manufacture them mostly in China. So we go to YKK in Japan, but we order the right zippers from their Chinese plants. Then we determine that, because of quotas and labour conditions, the best place to make the garments is Thailand. So we ship everything from there. And because the customer needs quick delivery, we may divide the order across five factories in Thailand. Effectively, we are customising the value chain to best meet the customer's needs. Five weeks after we have received the order, 10,000 garments arrive on the shelves in Europe, all looking like they came from one factory.

Whilst receiving orders from buyers, the large multinational manufacturers in Tier 1 frequently subcontract out these orders to smaller subsidiaries as well as to other factories that are harder to trace.

In numerous cases, the subcontracting is illegal in that the buying firms are unaware that their contractor has subcontracted out part of their order. Below Tier 1 the relationship between different levels of the supply chain alters radically, with a sharp increase in downward pressure in relation to price and turnaround times. This pressure pushes down through the different tiers in the chain to medium and small units, and to homeworkers. There are so many firms competing for business at lower levels that employers are willing to take on work that is badly paid. The further down the chain the work goes, the greater the pressures, bringing associated problems of excessive overtime and sub-minimum wages. These differences in the experiences of workers at different tiers of the supply chain are explored more fully in Chapter 5.

In order to illustrate the complexity of these networks we have used information from the research to build up a picture of the Gap supply chain (see Box 2.1). Gap was not a specific focus of the research, but many researchers found that they consistently met with workers who were part of Gap supply chains. The information supplied by workers

---

Box 2.1   Gap supply chain

Like many large retailers, Gap has regional and national sourcing offices. Its Asian Regional Sourcing Office is based in Singapore and there are national sourcing offices in key countries, including India, Pakistan, the Philippines and Bangladesh. An order comes through the regional sourcing office and is allocated to a national sourcing office. The national office passes the order on to one of the large manufacturers in Tier 1 with which it works, and that manufacturer is then the primary contractor.

In our example, the Tier 1 manufacturer is Blue Textile and Garment Manufacturing (see Figure 2.4). There are many manufacturers supplying Gap. However, according to employees working in Gap International Sourcing, the company tries to build up long-term relationships with 10–20 large manufacturers, depending on the country. These manufacturers are often multinational companies that have textile and manufacturing factories across the world. It is easier for Gap to work with companies that also produce textiles because it is cheaper and reduces the production turnaround time, not least because the company can co-ordinate their schedules so that production can be planned more efficiently.

Although Blue Garments produces textiles, it does not have the capacity to supply all Gap's needs, so Gap also orders textiles from large mills that do not have manufacturing capacity, represented in our diagram by Orange Textiles. When the textiles are ready they are sent to Blue Garments to be made into clothes.

Although Gap does not like factories to subcontract work out to smaller factories, this does happen. Blue Garments may send work out to (a) subsidiaries, (b) independent manufacturers and (c) agents. In the diagram these are 'Blue medium factory', 'Purple small factory' and 'agent' respectively. In some cases, Blue Garments will complete all the work for Gap in its own factory, but subcontract work for other brand names to Tier 2 manufacturers. Each Tier 2 manufacturer may then subcontract out to even smaller manufacturers, or home-workers, producing long, complex supply chains that Blue Garments know very little about.

When each manufacturer has finished its quota, it sends the finished garments back to the factory that subcontracted the work to it. All finished garments eventually come back to Blue Garments to be distributed to the stores. In the largest Tier 1 companies, distribution is done in-house: in some cases the company has a department that co-ordinates freight and distribution and, in other cases, large manufacturers like Blue Textiles and Garments have logistics companies as subsidiaries to which they subcontract the work.

The finished goods may be sent to the Gap national office or the regional office but it is more usual for Blue Textiles and Garments to send the garments straight to Gap's regional distribution centres: Gap-USA, Gap-Canada, Gap-Europe and Gap-Japan, from where they are shipped to the stores. The benefit for Gap is that Blue Textiles and Garments must pay the price of transport, distribution and administration, thereby saving Gap time and money.

It is not uncommon to find Gap clothes for sale in the department stores, malls and flea markets of the country where they were manufactured. This happens when too many garments are produced, an order is cancelled or the garments did not pass quality control. As these clothes are sold very cheaply, they force down the prices of clothes that are made for the domestic market, creating additional challenges for the smaller manufacturers that normally supply the domestic market.

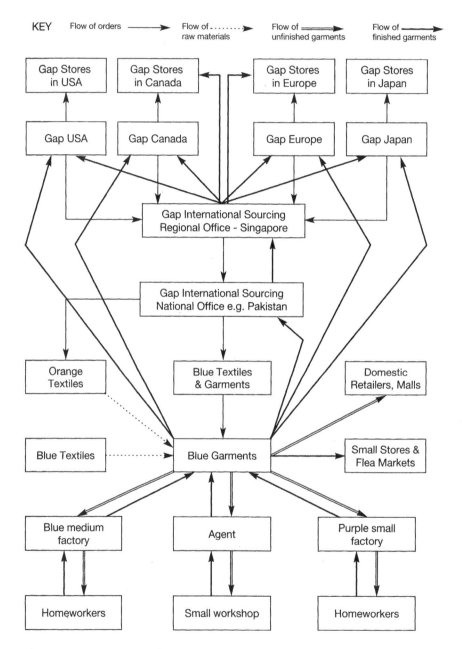

Figure 2.4 A Gap supply chain

was used to build up a picture of the more hidden levels of the chain. The names of companies in the chain have been changed as the purpose is to provide a model of how a garment industry supply chain operates in practice rather than to expose particular firms.

Basic Trends in Supply Chain Management

The garment retail industry can be broadly divided into seven segments: designer products at the high-fashion end of the market (Gucci, Dolce & Gabbana, Armani), top-quality high-priced brands (Burberry's, Diesel, Boss, Nike, Adidas) specialty stores with brand names (Rohan, a Children's Place, JJB Sports), mass merchandising (Nike, Adidas, Wrangler, Levis), discount chains (Wal-Mart, K-Mart), small retailers and the rapidly growing used or second-hand clothing market. These different segments share some common sourcing trends, which have helped to create a global production hierarchy. This hierarchy is based on closeness to the market and the value added to the garments, resulting in the geographic clustering of higher- and lower-value-added production in different locations.

In terms of volume, 'basic' clothing, which is typically sold through discount stores and specialty stores at the lower end of the market, accounts for nearly half of all garments sold. Demand fluctuates very little so the retailers source these high-volume products in distant low-labour-cost countries, and they are prepared to endure long lead times, particularly where low-cost transportation is used. In these segments the typical sourcing scenario might entail high-volume use of CMT (cut, make and trim) factories, where the value added is very low. This has resulted in the clustering of low-value-added manufacturing in South Asia (Gopal 2002) and, increasingly, in Central America and parts of Southern Africa (de Haan and Phillips 2002). As a result of these regions being confined to producing goods for export to the mass merchandising and discount markets, the garment industry is not proving to be a lynchpin industry in their economic development.

In contrast, higher-value-added production has tended to be concentrated nearer major markets. In the higher-profit retail segments dominated by the designer and speciality brand names, 'fashion-for-wardness'—that is the degree to which garments follow the latest trends and fashions—has become a prominent feature. This has meant a distinct shortening of the product life cycle and a proliferation of garment types which, in turn, has created increased demand uncertainty in the textile, garment and retail industry. Consequently, many of the manu-facturers producing higher-value-added garments are located in or near developed countries.

All segments of the garment industry are highly competitive and in order to maintain their position in the market, retailers and merchan-disers, particularly in the fashion and sportswear segments, have been engaging in three key practices: lean retailing, e-commerce and co-sourcing, all of which increase pressure down the supply chain.

Lean retailing

To avoid the risk of carrying inventory of increasingly unpredictable items, some companies place advance orders for small quantities of each garment and replenish stocks regularly, in some cases on a weekly basis. In order to operate in this manner, near vertical control from production to distribution with the elimination of middlemen and wholesalers, has been taking place. Such vertical networks can now be found in Wal-Mart, Uniqlo, Mango, H & M, C&A and Zara (for more on Zara, see Chapter 5, this volume). This reflects the streamlining of business struc-tures, discussed earlier in the chapter. However, rather than shortening their supply chains by concentrating on key manufacturers, these com-panies are creating vertical networks that extend beyond the first manu-facturing tier.

As an example, Zara, the fashion subsidiary of the Spanish multi-national Inditex, is viewed very much as the industry pioneer. Zara rotates stock 5 times a year leading to faster cycle times, which places considerable demand on its suppliers and their workers. Such a policy inevitably has three key employment impacts.

Firstly, suppliers need to be close to fabric and trim supply, which may result in the relocation of production, which in turn undermines job security. Historically, Inditex had a policy of sourcing primarily from within Spain using 24 different manufacturing subsidiaries. But more recently it has begun to broaden its supply base as part of its global retail expansion plans and now sources from over 2000 factories in Europe, Asia and Latin America (Lyne 2002; see Chapter 5, this volume). Sec-ondly, 'just in time' ordering inevitably begets 'just in time' production. Factories may be informed about quantity adjustments on the day of delivery, overtime may be immediately demanded and forced on workers (Raworth 2004:48-55). And, thirdly, in the mass discount market, sourcing companies may switch suppliers from one season to the next, looking for the best quality at the lowest prices. Suppliers deal with such unpredictability by opting for a flexible workforce, which is predominantly female and characterised by casualised work on a variety of short-term contracts. In larger factories, casual staff work alongside permanent staff, but have little or no access to whatever social benefits the permanent full-time staff have secured. This then creates additional tensions within the workforce (Dhanarajan 2004; and for more information, see Chapter 5, this volume).

E-commerce

With the advent of the Internet, apparel buying and selling has started to move on line with a proliferation of B2C (business to customer) and B2B (business to business) supplier networks (Hammond and Kohler 2000). B2B is having a significant impact on the supply chain. In Germany, for example, where 30% of all garment retailers and manufacturers engage in e-purchasing, under the new technique of RFP/RFQ (request for price/quote) buyers invite suppliers to bid on line for contracts, thereby generating downward pressure on prices which fail to reflect real labour costs (Ethical Trading Initiative 2003:51).

By 2000, a number of major retailers were already collaborating in on-line B2B exchange networks. As an example, in March 2000, 17 international retailers founded the World Wide Retail Exchange (WWRE) to enable participating retailers and manufacturers to simplify, rationalise, and automate supply chain processes. Currently the WWRE represents 64 companies including: Auchan, C&A Europe, El Corte Ingles, Galeries Lafayette, Gap Inc., J C Penney, Kingfisher, Kmart Corporation, Marks & Spencer, Meijer Inc., Otto Versand, Target Corporation, Tesco and Woolworths (Hammond and Kohler 2000:17). In such a climate of concentrated buying power, it is unsurprising that FOB (free on board[1]) prices for garments have been falling generally since 2000 (Clothesource 2003).

Co-sourcing

A further development in supply chain management is the emergence of what can be described as co-sourcing. Here the prime contractor, for example Levis, specifies on quality grounds the exact type and brand of fabric and/or components, such as thread or zips to be used by the subcontractor (De Coster 2004). This practice has had a major impact on jobs within the textile sector. As weavers and component manufacturers seek to negotiate global contracts for their products with garment manufacturers/merchandisers, they are compelled to ensure that their product follows the manufacturing locations from which these companies source their product. Coats plc, for example, a UK-based supplier of Levi Strauss and Nike, has developed a strategy of pursuing 'global key accounts' with retailers and merchandisers in an effort to ensure that their thread is a required component of any garment manufactured on behalf of those clients. Having progressively divested themselves of their last remaining clothing subsidiaries, Coats plc has opted to focus solely on the production of thread and zips and reorganised its production into bulk units in Hungary, Romania, China, India and Brazil, which are located near garment assembly zones. To meet demand and cope with market fluctuations, the company has also developed smaller units to provide customer service in local markets with shorter runs with specialist colours (Coats 2002). Understandably, this has had major implications for the relocation strategy within the company, impacting particularly severely on West European manufacturing facilities and their workers.

All these recent developments serve to increase the downward pressure that buyer-driven garment supply chains exert on employment terms and conditions in the industry. There is ceaseless pressure on costs and turnaround time and these pressures are felt most acutely by more than 11 million clothing workers who work for supplier firms (International Labour Organisation 2000:13). It is no accident that the major centres of garment manufacture are located in those parts of the world where wage costs are lowest, as Table 2.2 indicates.

Changes in the Regulation of Trade and Investment

The pressure on retailers and merchandisers to pursue the holy grail of high-quality, low-wage, 'one stop shop' sources of product is currently fuelled by the changing trade agenda, including the establishment of preferential trading and investment conditions under bilateral and regional trade negotiations in different parts of the world. And we explore the particular relationship between multinational capital, the global trade regime and the processes involved in foreign direct investment (FDI) in developing and less developed countries in the garment industry below.

Table 2.2 Hourly wage rates for selected countries, 2002

Region or country	Apparel industry
	US Dollars
East Asia:	
China	$0.68[1]/$0.88
Hong Kong	[2]
Korea	[2]
Taiwan	[2]
South Asia:	
Bangladesh	0.39
India	0.38
Pakistan	0.41
Sri Lanka	0.48
ASEAN countries:	
Indonesia	0.27
Malaysia	1.41
Philippines	0.76
Thailand	0.91
Mexico	2.45
CBERA countries:	
Costa Rica	2.70
Dominican Republic	1.65
El Salvador	1.58
Guatemala	1.49
Haiti	0.49
Honduras	1.48
Nicaragua	0.92
Sub-Saharan Africa:	
Kenya	0.38
Madagascar	0.33
Mauritius	1.25
South Africa	1.38
Andean countries:	
Colombia	0.98
Peru	[2]
Other countries:	
Egypt	0.77
Israel	[2]
Jordan	0.81
Turkey	[2]

(1) Reflects labour compensation for factories in China producing
      moderate to better apparel.
(2) Not available

Source: Jassin-O'Rourke (2002)

As global trade in textiles and apparel has increased, a complex and unique regime has emerged for managing the political and economic problems associated with increasing international competition. During the 1960s and 1970s excess capacity in production led to intense and difficult global competition as producers in the developed countries attempted to protect their markets from imports from low-wage countries through complicated quota rules and high tariffs, while developing countries responded with efforts to protect their industry using import substitution measures and bans. Eventually bilateral, country-by-country, trade policies began to emerge which culminated in the Multi-Fibre Agreement (MFA) of 1974. This agreement ratified countries' rights to impose quotas on trade within this sector, limiting trade in categories of apparel and textiles imports between countries. This was intended to be a temporary measure, to give rich countries time to restructure their industries before opening them up to competition from those countries with low-wage comparative advantage. In practice, however, the MFA became a major driver in the shifting geography of the sector, since the existence of quotas represented a de facto carving up of global manufacturing potential across the countries of the world. To tackle the inequalities of this quota system and the way in which it tended to protect production in developed countries, an international agreement, known as the Agreement on Textiles and Clothing (ATC) (for further information, see Chapter 9, this volume), was signed at the Uruguay round of trade talks in 1994, committing signatories to phasing out quotas according to an agreed timetable with a final date set for December 2004.

Since 1994 the restructuring of the textile and clothing industries of the US and the EU has continued apace. In the US, the American Apparel and Footwear Association estimates that 89% of clothing sales are from imports. The Japan Textile Importers Association now estimates that 87% of clothes on sale in Japan are now imported (Flanagan 2003). Production has continued to migrate to low-cost offshore locations in Asia, Africa and Central and Latin America, and there have been major jobs shake-ups in many parts of the world. The US clothing and textile industry, for example, has lost 316,000 jobs since 2001 (Barrie 2003:8). Similarly, the European textile and clothing sector lost nearly a million jobs between 1990 and 2000, and the challenges on the horizon make further job losses highly probable (European Commission 2003). Whilst thousands of jobs have been lost in the sector in the 'western' economies, similar tendencies have appeared and are accelerating in the industries of those countries to which apparel production originally migrated.

In 2003, Mexico lost over 100,000 jobs in the sector as merchandisers and manufacturers decided to switch production to China (Kearney 2003a; see Chapter 7, this volume).

A glance at the detail of the MFA and the ATC would lead one to conclude that the global trade in garments is heavily regulated, but this masks the way in which capital in the sector is able to pursue the maximisation of profit in an unfettered way. In practice, trade agreements have been used to cement the position of US- and EU-headquartered multinationals in the hierarchy of the value chain in the industry. The negotiation of bilateral trade agreements between national governments and the US and the EU, alongside the expansion of export processing zones (EPZs) to create cheap and attractive locations for multinationals 'regime shopping' in the sector, have increased the power of those at the top. In EPZs, the enforcement of national labor laws is lax and, in some cases, the outright repression of worker organisation is promised (International Confederation of Free Trade Unions 2003; see Box 2.2).

Box 2.2 Export processing zones

Export processing zones (EPZs) are one of the most controversial features of the globalising economy. Known by several different names—for example, free trade zones, *maquiladora* in Central America and special economic zones in China—EPZs share a set of distinctive features. Formally, EPZs are sites where imported materials can be worked on and then re-exported without incurring the usual taxes and duties. This formalistic definition, however, obscures many of the more pernicious aspects often associated with EPZs, including poor working conditions and trade union repression. The latter is widespread as the battle to attract multinationals to EPZs has led many host countries to suspend or curtail not just customs and tax regulations, but also employment rights, including the right to organise.

Although EPZs are largely a developing-world phenomenon, the modern export processing zone was actually pioneered at Shannon airport in Ireland. In 1960, the airport, which had been threatened by the advent of the transatlantic jet, was declared a tax-free production zone for value-added goods. This was seen by the government to be a cheap means of creating jobs. Here, as in the EPZs that emerged later, the zone was physically demarcated by high fences. Ostensibly, the

(Continued)

Box 2.2   (*Continued*)

purpose of the fences was to prevent tax-free goods from being smuggled into the regular economy. However, as Naomi Klein (2000) has argued, the fences also came to play an important role in maintaining control over the workforce.

EPZs emerged during the 1960s and 1970s as a key economic development strategy of many developing world governments and international organisations such as the IMF, World Bank and UN. That said, in 1970 only ten countries had established EPZs. By 1986, however, there were over 175 such zones across 50 countries and by the mid 1990s over half of the world's countries had EPZs. According to the latest International Labour Organisation figures, 42 million workers in 106 countries are employed in EPZs (International Labour Organisation 2003), with 30 million in China alone. Depending on the country, between 60% and 90% of these workers are women, most of whom are under the age of thirty.

Textile, clothing and footwear production, along with electronics, is the dominant activity in most EPZs. This is for several reasons. Firstly, EPZs are attractive to multinationals involved in textile, clothing, and footwear production as this is a very labour-intensive sector and, therefore, the quest for a large supply of 'cheap' labour is seen as critical to profitability. Secondly, the relocation to EPZs has been aided by the fact that textile, clothing, and footwear production is not very place-bound by virtue of its low capital intensity and as workers can be trained 'on the job.' Finally, production in EPZs has been enabled by the fact that distance of production from the end market is not such an important factor since most textile, clothing and footwear goods are relatively light and can, therefore, be freighted at minimal cost.

Source: Jeremy Anderson and Eva Neitzert, drawing on International Labour Organisation (2003).

These denials of basic worker rights are combined with a series of sweeteners to foreign multinationals, whereby the host countries make no demands for majority ownership, local content requirements, or the transfer of expertise and knowledge (Foo and Bas 2003). Box 2.3 illustrates the impact of these developments in the case of Ramatex and the Namibian government.

Box 2.3   The case of Ramatex in Namibia

In Namibia, in what has been seen as the most spectacular foreign investment in the country since its independence, the Malaysian multinational Ramatex opened a massive fully integrated operation to supply the European Union, the Middle East and the east coast of the US (under the terms of Africa Growth and Opportunity Act) in 2002. Drawing in the parastatals providing water and electricity (Namwater and Nampower), as well as the Windhoek municipality, the Namibian government put together an incentive package which included subsidised water and electricity, a 99-year tax exemption on land use, as well as over N$ 100 million to prepare the site including the setting up of electricity, water and sewage infrastructure. This was justified on the grounds that the company would create 3000–5000 jobs during the first two years and another 2000 jobs in the following two years. Namibia was also a particularly attractive location given the absence of any minimum wage legislation and an exemption for any EPZ firms from the provisions of the Labour Act of 1992.

Even before Ramatex started production, concerns were raised regarding the environmental impact of the company's operations, and tensions arose in relation to the discriminatory nature of the company's selection criteria. In the first full year of operation (2003), workers went on strike over pay, public transport and conditions. An operator's starting wage was N$1.50, or 12p an hour rising to N$3 (24p) an hour, and several hundred were suspended and are still awaiting reinstatement despite the successful outcome of an industrial tribunal case (Namibian 2003). Early in 2004, the Chinese foreign workers (there are some 2000 Chinese and Filipinos in the 8500 strong workforce) downed tools in protest at canteen hygiene, payments for company-imposed medical check-ups and failure to grant leave, and when the Filipino workers petitioned their embassy to come and inspect their conditions, the company refused the Consul General access to the site.

Source: Namibian (2004)

Just as multinationals have sought to protect their market share by entering into co-sourcing agreements with primary contractors, whole national textile sectors have lobbied to protect their market share via so-called preferential trade agreements and legislation. Examples here are the US African Growth and Opportunity Act (AGOA) of 2000, which provides duty- and quota-free access to apparel from sub-Saharan Africa providing it is manufactured with US/African-made fabric[2]. Likewise, the outward processing trade (OPT) rules of the EU allow the tariff-free importation of goods made in Central and Eastern Europe and the Maghreb using fabric of European origin (see Begg et al 2003).

With the expiry of all quota restrictions in the market for apparel and textiles on 31 December 2004, free trade in the sector is intended to prevail from 2005 onwards. Since World Trade Organisation (WTO) orthodoxy requires a free market there is also pressure to remove tariffs (ie, taxes on imported textiles and apparel) too. There are vastly different estimates as to the benefits that the phase-out of quota is supposed to yield. The Organisation for Economic Co-operation and Development (OECD), for example, predicts annual global welfare benefits ranging from $6.5 billion to $324 billion (Walkenhorst 2003). The views of major exporting countries are somewhat different however. A coalition of 71 apparel and textile trade associations from 38 countries is so concerned about the impact of the ATC that, in the so-called Istanbul declaration[3], it has called for a three year delay to the elimination of quotas.

Much of the talk in the garment industry has been about national 'winners' and 'losers' post 2005 (Anson 2003). While the critical factor in this competition would appear to be the availability of good textile infrastructure in national manufacturing bases, enabling suppliers to offer 'one stop shops', the decisions as to 'which workers in which countries will lose out and which will benefit from the quota phase-out is left entirely to multinational apparel corporations, whose only concern is the bottom line' (Foo and Bas 2003:9). In the end, the real winners are, as ever, the shareholders of the major multinational retailers and merchandisers in the global garment industry. Against a backdrop of such rapid and monumental change one thing looks set *not* to change and that is any prospect of decent work for the millions of women clothing workers in the sector (see Chapter 9, this volume).

Conclusion

The aim of this chapter has been to identify the major issues affecting the garment industry at the current time in order to provide the context in which to locate the research findings presented in the rest of this book. The chapter has explored the major trends identified in the garment industry, and it provides a bridge linking theory and reality, global trends and the daily lives of workers.

The global garment industry is undoubtedly going through a period of intense change, both structurally and politically. This can been seen in the supply chains and networks in the industry. There is consolidation of power at the top of the industry with closer relationships between retailer/marketers and key manufacturers, and increasingly complex and multifarious relationships at the bottom of the chain. The operation of these chains is constantly altering with the adoption of new strategies such as lean manufacturing and e-commerce. Fundamental changes in international regulation will also affect the industry when the MFA is finally phased out sometime after 2005.

Yet behind all this change there is a paradox, for these changes are making little real difference to actors who are not already in positions of power. Powerful actors are able to use these changes to enhance their positions. Those in less powerful positions, whether they are developing nations, small manufacturers or workers, find that their situations remain the same or more fragile than ever. Constant downward pressure, particularly on prices and turnaround times, continues to be a salient features of the sector. Yet, as the following chapters illustrate, this situation is being contested and in identifying and exploring other avenues for change, we join a global discussion about how to challenge and reconfigure power relationships within the industry for the long term.

Notes

1   Free on Board refers to the price charged for a product by a supplier. The price does not include delivery and insurance for the goods.
2   AGOA does provide for fabric from so-called 'third' countries to be used in garments exported to the US from less developed countries.
3   Cf. www.apparelresources.com/defaultnextseven.asp?msg=4504&cod= newsindetail&nam=

3

Organising and Networking in Support of Garment Workers: Why We Researched Subcontracting Chains

Angela Hale

Introduction

The research in this book has emerged from Women Working Worldwide's involvement in what can be seen as a new form of labour internationalism. The plight of workers in globalised industries, and particularly the garment industry, has drawn in new constituencies of support both locally and internationally, which have developed sometimes uneasy alliances with the established trade union movement. Together they have pushed for new concepts of corporate responsibility, which are beginning to produce some benefits for workers. It is interesting that it is poor women workers who are at the centre of this movement. Most workers in global manufacturing industries such as garments are young, often migrant, women who are employed on a short-term, casual basis—the very workers who have been marginal to traditional trade union organising. The imagination and determination of these workers has combined with the vision and commitment of organisers and campaigners within both producing and consuming nations to challenge the exploitative nature of globalised production.

Collaborative projects such as the action research reported in this book are part of this new labour movement. The research proposal developed out of the need of both local worker-based organisations and international campaigners to understand more about how the global garment industry operates. It also developed out of a recognition of the need to build organising links between workers at different points in supply chains and between workers and consumers. The findings of the research are being fed back into the work of these different organisations and are also being used to build a wider constituency of support (for further information, see Chapter 4, this volume).

This chapter outlines how this new movement developed and the ways in which it has focused on the garment industry. The starting point is the self-organisation of women workers, beginning with workers in the export processing zones, which expanded rapidly in Asia in the 1980s. Over the past twenty years many local organisations have emerged supporting the rights of garment workers who are beyond the reach of trade unions or are working in situations where trade unionism is suppressed. Since trade unions only represent an estimated 5% of garment workers worldwide, these organisations have become important players, establishing links with wider international networks, including consumer-based organisations in Europe and North America. This chapter will look at how these international networks operate, how they link trade unionism to other organising strategies, and how their exposure of the abuse of women workers in the garment industry has prompted a response from companies in the form of 'corporate social responsibility.' The account focuses on links into Europe and will be presented from the point of view of Women Working Worldwide (WWW), which has been networking with women workers' initiatives since the mid 1980s. It will include an explanation of how this research on garment industry subcontracting emerged from WWW's previous work with partner organisations in Asia, notably an education and consultation programme on company codes of conduct. Finally, there will be a consideration of how the research can be used to develop new alliances and to further challenge the negative impacts of globalised production.

Women Garment Workers and the Emergence of New Forms of Organising

Women as the new global labour force

The massive increase in the employment of young women in export production has been one of the most striking aspects of the integration of poor countries into a globalised economy. Nowhere is this more apparent than in the garment industry. The clothing industry together with textiles and footwear provides the largest amount of manufacturing employment for women globally and is one of the most widely dispersed industries throughout the world (International Labour Organisation 2000). It was one of the first industries to relocate manufacturing processes to Asia and elsewhere in the 1970s, when improvements in transport and communication made geographical distance less significant. This was also a time when export production was being strongly advocated for poor countries, by agencies such as the World Bank, as a stimulus to their industrial development. What emerged, of course, was not industrial development but, as Fröbel et al (1980) point out, the selection of particular sites by multinational companies for the relocation of labour-intensive industrial processes. The result was a new 'international division of labour', with workers in different parts of the globe involved in the same production line (Fröbel et al 1980). It was feminist economists, notably Elson and Pearson (1981), who drew attention to the gendered nature of this division of labour. They pointed out that the profitability of the textile and garment industry has always been premised on the exploitation of female labour. The reasons given for the preference for women workers in export production in Asia were remarkably similar to those given in Europe, reflecting patriarchal structures in which women are not expected to earn a living wage and are regarded as pliable and docile (Elson and Pearson 1981:22).

What soon became apparent was that this new industrial labour force of young women were far from docile. Most were sent from rural areas by their families in order to supplement the family income, and it was the first time they had earned an income. Their experience of the exploitative conditions in export factories led them to begin to organise and demand their rights. This organisation often took place in the face of heavy repression, involving both state and employers. The Asian countries to which the garment industry was being relocated were authoritarian regimes that restricted the development of effective trade union movements, initially Korea, Taiwan, Singapore, and later Sri Lanka, Indonesia and then Bangladesh. Many of the factories were opened in newly created special industrial zones, known as export processing zones (EPZs), free trade zones (FTZs), or *maquilas* in the case of Central America. These provided exemption from normal labour legislation, which meant that even countries with good labour legislation did not have to implement this within the zones. This is still the case today in that trade union organising is extremely difficult, if not illegal, in most

EPZs in Asia and elsewhere. These zones are geographical enclaves surrounded by high fences with armed security guards who control access and are often used to discipline workers and prevent industrial action (International Confederation of Free Trade Unions 1996; see also Chapter 2, this volume). This is how a worker in the Philippines described her arrival at work in 1985:

> Soldiers dressed in combat fatigues and well armed with automatic weapons cast their eyes over the bus before waving us through... The main industrial complex is surrounded by high walls and a wire fences. All workers have to carry identity cards and have to queue while there are security checks... This industrial zone is governed by its own armed police force with its own intelligence service and network of spies. (Donald and Bamford 1990:9)

It was in response to the situation of EPZ workers that Women Working Worldwide first began. We came together in 1983 as a group of women activists from research and development agencies in the UK who had made contact with some of the Asian activists and academics who were beginning to document working conditions in EPZs. A conference was organised entitled 'Women Working Worldwide: The International Division of Labour in the Electronic, Clothing and Textile Industries,' and representatives were invited from Sri Lanka and the Philippines to provide testimonies from workers. They gave disturbing accounts, for example of extreme work pressures involving forced overtime. Padmini, a worker from Sri Lanka, reported: 'When we have to work a night shift, it means that we work a day shift, then a night shift, then another day shift... altogether we are there for 2 days. We have a short break for dinner and then between 2–3 in the morning we have another break, but we try not to sleep because then it is difficult to work again' (Rosa 1983:5). We were told that many workers developed illnesses, which, together with the low pay, meant that the only way they could fulfil their family obligations was by denying themselves adequate food. However, in spite of these deprivations, participants also reported that workers were beginning to find ways of resisting this exploitation.

Although the spread of EPZs was one of the most dramatic developments of the 1980s, the globalisation of the garment industry has in fact relied even more on the labour of women workers in small urban factories and workshops. In Bangladesh, for example, the majority of garment production has taken place in small factories in the capital city of Dhaka. The garment industry expanded rapidly here during the second half of the 1980s as production was being relocated from the higher-wage economies of East Asia. Thousands of young women were sent from the countryside to work in these factories. Most were in apartment blocks and other buildings where the upper floors have been converted to makeshift factories. Home thus becomes the workplace for many garment workers and these women are the most marginalised and underpaid workers. This dispersed workforce has been beyond the reach of the official trade union movement, but, despite this, workers have developed their own ways of organising.

The following section provides an account of how garment workers in a variety of different locations have developed new ways of organising as women workers in a globalised industry. This includes not only local workplace resistance but also the development of regional and international networking. The account is given from the perspective of WWW, which, following the 1983 conference, was established initially as a solidarity group, supporting this emerging women workers movement and raising awareness of their demands in the UK.

Organising in Asia

One of the first examples of organised resistance by women garment workers took place in Korea in the 1980s. The Government of Korea had adopted a policy of export oriented industrialisation, and at the same time, had introduced new laws banning trade unionism as one means of attracting foreign investment. There was a massive expansion of garment production and thousands of young women migrated to industrial areas to take up jobs on the production lines. Much has been made of how the Korean propensity for hard work contributed to the rapid development of the country's economy, but the other side of the story was the systematic abuse of women workers' rights. Despite this, women began to organise and then take to the streets in massive demonstrations to demand the right to form trade unions (see Box 3.1). This right was subsequently established, but by that time many women factory workers had lost their jobs as the companies began outsourcing production to other low-wage economies in Asia and elsewhere.

As WWW developed links with EPZ workers in the 1980s, we learnt of many other examples of collective organisation by garment workers. The potential for the self-organisation of women workers in EPZs was also being noted by feminist economists such as Elson and Pearson (1981). They argued that the subordination of women as a gender

---

Box 3.1   Organising garment workers in Korea

'In the 1970s the workers movement consisted mostly of women...
Many university students stopped study and went to the companies to
work with the workers, and they raised the consciousness of the
workers to fight for their rights... Government policy was very strict.
Even if two or three women tried to organise a group this was against
the law and you could be put in jail. Organisers were arrested or
beaten by gangsters. But even in this kind of situation they did not
give up their struggle... The tactics they used were to go to the
women workers houses in the evening when work was finished at
8 or 9 o'clock and have a meeting around 10... Most of the workers
were very young and unmarried and even though they were working
a twelve hour day they would still come. They would write by candle
light for secrecy and discuss how to improve their conditions.'

Source: Maria Rhei (Women Working Worldwide 1998:11)

---

meant that the new female workforce were particularly skilled at presenting themselves as docile and subservient, whilst simultaneously
building solidarity as workers. They gave an example from behaviour
observed by Heyzer (1978) in a factory in Singapore: 'Here women
workers were always on guard when the supervisors were around
and displayed a characteristic subservience; but in the absence of supervisors behaviour changed. Far from displaying respectful subservience,
workers mocked the supervisors and ridiculed them' (Elson and Pearson
1981:27). Elson and Pearson also noted that most workers were away
from the restrictions of home for the first time and were not only
working but also living in close proximity. Most were in boarding
houses outside the EPZs with as many as twenty sleeping in the same
room. In spite of the long hours of work they were able to spend time
together discussing their lives in ways that had previously been impossible. As Kumudini Rosa reported, this provided the basis for the emergence of new 'spontaneous and innovative forms of organising' (Rosa
1994:85).

These innovative forms of organising included the use of 'eye contact'
in factories in Sri Lanka, where workers were not allowed to speak, and
the use of local language in the presence of foreign management. As
one worker reported: 'We have our own ways to organise ourselves.
This is very important for us. After a period the workers have all got

accustomed to these methods' (Rosa 1994:86). In the Bataan EPZ in the Philippines, women workers told us how they organised mock funerals of particularly exploitative factories and developed the practice of all wearing company T-shirts inside out when it was agreed that action was needed on a particular issue. Looking back ten years later, Elson and Pearson (1998) note the significance of these 'innovative strategies' which were quite alien to the established trade union movement.

These initial worker activities soon led to the formation of local organisations such as women's centres based outside the EPZs. Since workers were not allowed to organise as trade unions, most operated under the cover of religious or community organisations. Two such organisations, Da Bindu and the Women's Centre, were set up in Sri Lanka in the 1980s. The women were initially linked to Christian organisations, but soon became autonomous. Da Bindu began publishing a newsletter for and about the workers, which provided useful information and also recounted personal stories. As they put it: 'We went to the boarding homes and handed out these publications. As we made progress in distributing information to workers they saw they could talk to us about their problems' (Women Working Worldwide 2000b:34). The newsletter is still being printed and is translated into English for wider circulation. The stories illustrate not only the suffering of workers, but also their innovative and effective organising strategies. One story included during 1990 was entitled 'An eye is worth 5 dollars' and related to an incident where an entire workforce immediately and silently walked out of the factory and refused to go back until the injured worker was taken to hospital. Then they demonstrated outside until it was agreed that the supervisor involved would be dismissed (see Box 3.2).

Da Bindu and the Women's Centre continue to organise in support of women workers in EPZs in Sri Lanka (and were involved in the research reported later on in this book, see Chapter 4, this volume). In 1997, Da Bindu produced a report on the situation facing women workers who had been employed in the EPZ for more than 5 years. The abuse of workers' rights was clearly still widespread: 'The supervisors often assault and physically abuse women workers. They subject them to verbal abuse, stab them with pens, and subject them to innumerable indignities' (Da Bindu 2000:17). However they were also able to report a number of successes in relation to the rights of individual workers and disputes in particular factories. Perhaps the most significant achievement of these women workers' organisations has been to contribute to an increase in the consciousness and confidence of workers to

---

Box 3.2 Sri Lanka: An eye is worth 5 dollars

'I reported as usual to work on 22nd Sept 1990. I was ordered to sew shirts that day. While I was sewing a supervisor came from behind and shouted at me "Is this how you were asked to sew?" Whilst saying this he hit me on the head. My head struck the machine and an iron piece injured my left eye. The company told me to go home. With unbearable pain I went to the medical centre in the FTZ. The medical officer told me that he did not have either the equipment or medicine to treat me and that I should get myself admitted to the general hospital. I began walking to the main gates of the Zone and at the main entrance I lost consciousness. When I regained consciousness I was warded in Negombo Hospital.

It is difficult to talk. The pain in my eye is unbearable. I can't see anything with this left eye. I won't be able to work. My younger sister also works in the FTZ. Because she took 5 days leave to look after me, she was not paid her salary this month either. The management visited me and gave 200Rupees (US$5) and told me to buy what I need! But he also told me that the supervisor who hit me has been dismissed.'

This is written by a woman worker who was becoming a leader, often representing the demands of the workers in her section. When these atrocities happened workers of the factory demonstrated outside the factory gates and demanded the immediate dismissal of the supervisor. It was this pressure that forced the management to dispense with the services of the supervisor. The factory is a Hong-Kong Sri Lankan joint venture producing garments. Workers claim that around 4000 workers are employed in this enterprise.

Source: Da Bindu (1990:1)

---

demand their rights. This has enabled workers to begin organising as unions once a change in the law made this possible. In 1999 it became obligatory even for employers within EPZs to recognise trade unions if more that 40% of the workforce were members. Within weeks, the new FTZ Trade Union was registered, with branches attempting to establish themselves in factories all over the zones.

The organisations that emerged in the 1980s recognised the limitation of taking action only on particular cases and against immediate employers. They knew there was a need to confront the power of companies in Europe or North America who controlled the supply chain.

They therefore welcomed WWW's role in providing information about these companies and organising awareness-raising and campaigning activities in the UK. One early example of this collaboration was a campaign that WWW ran in 1989 in support of workers in a garment factory in the Philippines owned by a British company. These workers were involved in a growing movement in the Philippines that incorporated labour rights into a wider campaign against political and military repression. A wave of strike action in the early 1980s had built up strong organising potential amongst workers in the Bataan EPZ. Workers in this particular factory were therefore well prepared for a struggle. The story of what happened not only provides an early example of international collaboration, it also shows that, even at this time, companies used subcontracting to undermine effective organising. The account also illustrates how UK retailers responded to the abuse of workers' rights in the 1980s, before the development of the movement for corporate responsibility (see Box 3.3).

Box 3.3   The Philippines: Campaign in support of a year-long lock-out

On 20 September 1989 IGMC, a British company in the Bataan EPZ in the Philippines, locked out its 1000 workers and announced that it was closing down. The reason given was that (in spite of their healthy profits) they could not afford to pay the legal new minimum wage, which had been established in response to a general strike by the trade union movement. Within hours of the announcement the women workers began a night and day picket outside the factory, often bringing their children with them: 'With only makeshift cardboard shelters to protect them from the blazing sun and intermittent rain, they say they will not move until the factory reopens.'

The workers asked WWW and others to organise a boycott until the factory reopened and reinstated the workers. A loose campaign alliance was formed which included the North West Region of the Transport and General Workers Union. Approaches were made to the high-street shops selling garments with labels from the factory, but, with the exception of Littlewoods, campaigners were told that the pay and working conditions in the Philippines were not the concern of retailing companies here. When the organisers produced publicity leaflets and gave these to customers, a libel writ was issued by the parent company based in the UK.

> ## Box 3.3 (Continued)
>
> Meanwhile in the Philippines workers were still camped outside the factory. Six months after the lock-out it was reported that: 'The conditions of the workers are appalling. Many have lost their homes because they have been unable to keep paying the rent, so they no longer have any home except the picket'. The workers were staying even though it had become apparent that the company was not going to reopen the factory. The work had been taken elsewhere. Even before the lock-out the company had begun subcontracting to small factories outside the FTZ, finding places where workers were unorganised and it was possible to avoid paying the minimum wage. The workers continued the factory picket in order to claim back wages and benefits and, if necessary, their redundancy pay. Eventually, almost exactly one year after the initial lock-out, an agreement was reached through which the workers were able to claim all these rights. The workers insisted that the agreement also included a clause which stipulated that the company would not proceed with the legal cases against WWW and other UK campaigners. Solidarity worked both ways.
>
> *Source*: Appeal bulletins produced by WWW for the Campaign Group in October 1989 and March 1990

One common characteristic of the women's groups that have been set up in support of garment workers has been their recognition of the need to organise within the broader framework of workers' daily lives as women. For example, Karmojibi Nari was set up by women trade unionists in Bangladesh in 1994 to address the specific needs of women garment workers. The aim is to not only encourage organisation amongst the workers but also to build their confidence as women and to lobby for equal rights at work. Similarly the Women Workers' Organisation in Pakistan is concerned not only with organising and educating women workers but also with the elimination of discrimination and the equal participation by women in all spheres of life. Both organisations work alongside the trade union movement but their priorities are often quite different (both organisations have been part of the research reported in this book; for more information, see Chapter 4, this volume). Their demands are based on a recognition that workers are being exploited not only because they are poor but also because they are women. As Elson and Pearson (1981:38) pointed out, 'Struggles arising from the development of world market factories will remain seriously deficient from the point of view of women workers if they deal only with economic questions of pay and working conditions, and fail to take up other problems which stem from the recomposition of new forms of subordination of women as a gender.' In garment factories in Bangladesh and elsewhere this has typically involved the substitution of paternal control in the home with control by male supervisors in the factory and the replacement of control by confinement with control through the threat of sexual and physical violence. Many garment workers prioritise these concerns and they are taken up as major issues by women workers' organisations (Women Working Worldwide 2001).

Since women workers' organisations tend to take the concerns of workers as the starting point, it is inevitable that the issues taken up relate not only to the situation within the workplace but also to the wider community. For example, personal safety is a concern that relates not only to safety in the factory but also to travelling to and from work. Da Bindu reports many incidents of women being assaulted when travelling to the FTZ in Sri Lanka, particularly since they often work late into the night. Organising strategies have included efforts to ensure that workers always travel together and that they all shout out if they think any one of them is in danger (Da Bindu 1990). Other issues may relate to the difficulties of actually getting to work. The Friends of Women (FOW) in Thailand organised a campaign to prevent the building of a road which would separate workers' homes from the factories where they work (Women Working Worldwide 2003b:27). FOW work with both unionised and non-unionised workers and they use various strategies to encourage women workers to organise. In the case of small garment 'shop-houses,' they may start by taking workers out on a trip on their day off so that they can begin to talk to each other, or they may suggest that they start a savings club, which could be a first step to forming a union.

The Self Employed Women's Association (SEWA) in India has demonstrated that, by taking this broader approach, an effective membership organisation can be built up even in the case of homeworkers. This has involved, for example, extending the notion of collective bargaining so that it relates not only to employers but also to local and central government. As an example, SEWA is currently lobbying for garment industry homeworkers to be paid the minimum wage and to be issued with ID cards so that they can prove they are workers (Unni and Bali 2002:140; see also Chapter 8, this volume). Many similar organisations have emerged in support of homeworkers, not only in poor countries but also in richer countries, such as the UK (see Chapter 6, this volume). Since many homeworkers switch from garment production to other activities, membership is not based on the particular sector but on their situation as workers. For example, Patamaba, an organisation which works with garment homeworkers in the Philippines, bases its organising on the overall right to a livelihood rather than issues relating to particular occupations (Ofreneo et al 2002).

Over the last decade, innovative strategies have also been developed in response to the increased informalisation of women's work. Many women previously employed in factories have been forced into working on poorer terms and conditions in small workshops or at home. This was illustrated most dramatically in Korea, when thousands of women workers were dismissed from regular jobs following the financial crisis. In 2000 the Korean Women Workers' Association responded by setting up the Korean Women's Trade Union which was 'organised to expand and enhance women's right to work and right to unite against the worsening realities confronting women workers such as the pervasiveness of "women fired first" and the rapid increase of irregular women workers' (Da Bindu 2000:19). Any woman can join the union whatever her place of work, and even if she becomes unemployed her membership is maintained. 80% of the members are 'irregular' workers, and they work in a wide range of sectors (Women Working Worldwide 2000b:19).

Organising in Central and Latin America

WWW's strongest links have always been with women workers in Asia. This is mainly because of supply chain connections, since most garment outsourcing from the UK and Europe has been to countries in Asia. However, links have been built to similar organising initiatives by women workers in industrial zones in Mexico and Central America that supply the North American market. In Mexico there was a surge of popular activity in the lead-up to the signing of NAFTA (the North America Free Trade Agreement) between the US, Mexico and Canada in 1994. This agreement opened the door to US investment in Mexico and there was a huge increase in the number of *maquilas* (EPZs) on the Mexican American border. A networking organisation, Mujer a Mujer (woman to woman), was set up across the three countries, and one of its main concerns was the impact of the agreement on women workers. In Mexico the network involved a women's support group called Factor X situated in Tijuana, just across the border.

Links between activists in Asia and those in Mexico and Central America were developed by WWW in association with Mujer a Mujer, and subsequently, the Maquila Solidarity Network in Canada. A representative of Factor X spoke at a WWW conference called 'World Trade is a Women's Issue' in 1996 and she described conditions very similar to those in Asia: 'Most of the workers are women between 16 and 24 years of age. A normal working day is 10–12 hours and the women have to live in shanty towns without sanitation and services. Many women experience sexual harassment' (Women Working Worldwide 1996:7). In 1998 WWW organised a workshop in association with the UK-based Central America Women's Network (CAWN) which brought together women workers and activists from Korean-owned factories in Central America and Asia. They described factory regimes and coping practices adopted by women workers that were almost identical. For example, there was a common limitation on the number of times women could go to the toilet. Martha Christina Gonzanlis, a worker from Guatemala, reported: 'Sometimes we have to queue up and if you are in there for over three minutes they write it down in their note books. If the number of times you go accumulates then they start to shout at you and you get hassled' (Women Working Worldwide 1998:13). In the workshop it was revealed that it was common practice for women in both regions to take plastic bags into work in case of emergencies.

At these workshops women activists have been able not only to share problems but also to organise strategies. At the same workshop in 1998, Sandra Ramos from Nicaragua talked about the establishment of an organisation called the Maria Elena Cuadra Women's Movement for Employed and Unemployed Women in 1993 (see Box 3.4).

As in Asia, there are other organisations that have been set up to support garment workers who are more isolated and less visible than the *maquila* workers. One of the earliest and most dramatic examples was the garment workers' union set up in Mexico following the earthquake on Sept 11th 1985. The earthquake struck the centre of Mexico City in the early hours of the morning before many people had come into the city. However, the women garment workers were already at work in the small inner-city sweatshops, and thousands of them were crushed in the debris. The horror of this event motivated surviving garment workers to form their own union (see Box 3.5).

Women workers in Latin America have also suffered from the consequences of informalisation in a manner similar to the situation in Korea

Box 3.4 The Maria Elena Cuadra Women's Movement (MEC) in Nicaragua

'I have been a trade unionist for 16 years but the trade unions don't take account of womens needs... The women in the maquila cannot keep waiting for the trade unionists to wake up one fine day and realise that there is a situation of discrimination and violation of women workers in the maquila plants. So a group of us decided to get together women working with the maquila in Central America, independent groups, to sit down and see what we could do together. We thought we needed different approaches, different paths forward, but the same objective, which is the rights of women workers. And we decided to set up our own code of ethics. We began our campaign, which lasted for a year, and everybody said "You are out of your mind, who on earth is going to be interested in this? You have to have someone who supports you," in other words some important man from the union. But for us the trade union is one path forward... It's not the badge that matters, its what you do, and you have to understand that times have changed, and what we have to do in these times is have a clear way to achieve our objective, which is to organise in the maquilas in the way workers want' (Sandra Ramos speaking in 1998).

In 1997 a mass meeting of maquila workers was organised where workers prioritised their demands and drew up their own code of practice listing the rights they wanted respected by employers. They then ran a campaign called 'Employment yes, but with dignity' and built up support amongst other workers and the community. On February 1st 1998, in front of 500 women workers, the Nicaraguan Minister of Labour signed the code. The next day the owners of 23 factories in the zones agreed to comply.

The MEC has also been instrumental in setting up a Central America Network in Solidarity with Women Workers in the Maquilas, which is monitoring the code and encouraging its adoption in other Central American countries.

Source: Women Working Worldwide (1998:15)

after the financial crisis of the 1990s. They have also responded with similar forms of organisation. Ana Clara, a group in Santiago, Chile, began working in 2000 with garment workers who have been forced to

Box 3.5 The September 19th women garment workers' union in Mexico

'It is a disgrace that it required scenes of bodies crushed under the debris and between the rolls of cloth and machinery or of insensitive owners trying to recover their machinery in the midst of the tears of garment workers' families, to raise awareness of such an important, yet so marginalised sector of society.

'... The workers were indignant at the indifference and negligence of both businessmen and politicians following the earthquake. More than 5000 garment workers united and organised themselves—first to demand the rescue of their fellow workers and afterwards to defend their rights against the injustices of the bosses, who had paid no compensation to any workers, living or dead. The workers, who all worked in isolated workshops recognised that each one of them had suffered the same forms of oppression and decided to organise themselves independently.

'... On 20th Oct 1985 these workers succeeded in being recognised as the National Union of Garment Workers, or the 19th Sept Union. This was a great achievement, for, since 1976, no democratic union had been legalised in opposition to the official unions. The union is unique because it is a democratic organisation run by women workers. Within the first four months of its existence the union succeeded in forcing 80 employers to compensate more than 8000 women workers who had lost their employment as a result of the earthquake. This achievement encouraged many other garment workers to affiliate to the union.

'... Whilst the general conditions of thousands of garment workers have benefited from the union's activities, new problems have arisen since it was formed. Recession forced many of the illegal workshops to close down. Ironically too, the union's very successes provoked retaliations which undermined its position. The union's defence of the workers led some owners to close or relocate factories... Despite coercion and manipulation, the 19th Sept Union has maintained an alternative form of organisation for women workers in the garment industry and has served as a point of reference for other women workers.'

Source: Tirado (1994:106)

leave factories and become employed on poorer terms and conditions in small workshops or at home. The first stage was to locate these workers. To do this, Ana Clara trained the homeworkers they knew to carry out surveys and to find other workers. On the basis of this education and training, homeworkers are now being encouraged to set up their own unions: 'In December 2001 the first national meeting of home-based workers was held in Santiago, and women started to talk about setting up their own organisations. In 2001 four local unions of homeworkers were registered and about a dozen local groups set up in other areas' (Homenet 2002:5). As with the initiatives in Asia, these activities in Central and Latin America demonstrate the ability of women-led organisations to respond to the challenges of an increasingly dispersed workforce.

Regional and international networking

Of course, WWW has not been the only organisation to develop links with these locally based initiatives and to support their work. A number of other networks were set up around the same time. The Committee for Asian Women (CAW) started as early as 1978 to bring together organisations representing women in both formal and informal workplaces throughout Asia with the statement that 'We do not see women workers in Asia as just workers but also as women' (Committee for Asian Women 2002:36). The Asia Monitor Resource Centre (AMRC) was set up in an office in Hong Kong, with the central aim of supporting 'a democratic and independent labour movements in Asia' (AMRC website). Later Transnationals Information Exchange (TIE) Asia began networking in response to 'the growing number of mostly women workers, who are largely unorganised and precariously employed in the export-oriented textiles, garment and related industries, within and outside the zones' (TIE website). Meanwhile, in 1994 Homenet was formed to provide a global network for the growing number of homeworker organisations.

What all these initiatives have shared is a commitment to working with grassroots workers' organisations, not only trade unions, but also the newly forming women workers' groups. Whilst the right to form trade unions has been maintained as a central goal, there has been an additional demand that this be done in a way that fully represents the needs of women workers. These new networks have also been able to develop flexible ways of working which are difficult for the international trade union movement. Like WWW, they have undertaken a wide variety of activities, ranging from support for local organising and specific workplace disputes to research, campaigning and advocacy work at an international level. All are small organisations, but the fact that they are strongly rooted in a workers' movement has sometimes meant that they have been able to exert a considerable influence on global decision-making. For example, it was largely due to the efforts of Homenet that an international convention on homeworking was adopted at the ILO Conference in 1996 (Homenet 1999).

Linking the Women Workers' Movement to the Drive for 'Corporate Responsibility'

Whilst new ways of organising were developing amongst women garment workers in Asia and elsewhere, an awareness was growing amongst consumers in richer countries that the proliferation of cheap fashions is based on the exploitation of labour in poorer countries. This resulted from the work of Northern-based groups, including WWW, who collaborated with journalists to expose the link between particular factories and popular retail outlets. One of the first such exposés in the UK was a World in Action documentary entitled *Rags to Riches* in 1984, focusing on factories in Thailand where young women were kept in day and night to supply UK outlets including Littlewoods, General Universal Stores and Debenhams. This documentary was used by WWW and others to campaign for retailers to be held responsible for conditions throughout their supply chains. This focus on retailers was a new departure for campaigners. At the time, anti-corporate campaigns were focused on the power of big producers such as oil companies, mining companies and food conglomerates, and it was a novel idea that retail outlets could also be major players in the global economy. Yet, in the case of the garments, it was becoming clear that the development of new technology was contributing to a shift in power from producer to retailer, and that it was the growing competition between these retailers for cheap and flexible production that was driving down labour conditions in the industry. As the Bataan EPZ example above shows, retailers were turning a blind eye to this reality and only acknowledging responsibility for their own direct employees. This needed to change. Retailers and brand-based companies needed to be made accountable for the conditions of all workers involved in the production of the garments they sell.

This section will start by looking at the beginnings of European campaigning that has targeted garment retailers. Through computer communications, these campaigns have developed links with similar campaigns in North America and Australia. Taken together, they can be seen as developing into a significant social movement based on the demand that all companies take responsibility for working conditions throughout their supply chains. The link between this movement and the development of 'corporate social responsibility' will be demonstrated through an account of WWW's involvement in setting up the UK Ethical Trading Initiative (ETI). Whilst the ETI and other similar initiatives are based on an acknowledgement of company responsibility, they remain top-down strategies, far removed from the everyday lives of workers. The following account will show how the focus of some of WWW's work has therefore been to try to bridge the gaps between these high-level initiatives and women workers themselves. In particular, it will look at the education and consultation programme on company codes of conduct, which WWW carried out with garment workers in Asia between 1998 and 2001. It was the awareness of company codes which developed from this programme, together with the increasing informalisation of work, which led to a recognition by women workers' organisations of the urgent need to understand more fully how international subcontracting actually operates.

European campaigns in support of garment workers

Whilst WWW was organising support for employees of the Intercontinental Garments Manufacturing Corporation (IGMC) in the UK, a research organisation in the Netherlands, SOMO, was beginning a campaign based on a book it had published on C&A (Smit 1989). One of C&A's suppliers was IGMC, and a link was established between WWW and SOMO, which can be seen as the beginning of European networking in support of garment workers (Shaw 1997). In 1990, SOMO launched the Clean Clothes Campaign (CCC), initially targeting C&A. Demonstrations were organised outside C&A shops and customers were asked to sign petitions in support of workers' rights. Later, the campaign was broadened to include all major garment retailers and support was received from the Dutch trade union federation, FNV, and from NOVIB, a development agency which is a member of Oxfam International. In 1995, with funding from the European Union, a new project was set up with the aim of developing the CCC throughout Europe. In the UK, WWW invited a range of groups concerned with the garment industry, including development agencies, church groups, fair trade outlets and solidarity groups, to form a network called the Labour Behind the Label as the UK arm of the CCC. Labour Behind the Label now has the support of British trade unions and has recently become established as an organisation in its own right. Similar networks now exist in most European countries, all part of the wider CCC (for more information, see www.cleanclothes.org).

The CCC has proved itself to be an extremely vocal and effective international campaign for workers' rights. This has only been possible through holding together a network of different organisations and individuals campaigning across a growing number of European countries. The CCC has a secretariat in Amsterdam, and regular meetings are held with country representatives. However, each national coalition develops

---

Box 3.6   Victory at Jaqalanka, Sri Lanka

In 1999 the legal situation in Sri Lanka changed so that it became obligatory for even employers within FTZs to recognise trade unions if more than 40% of the workforce were members. However many employers are resisting this change. One of these was the Jaqalanka factory in the Katanuyake. By 2003, 200 of the 400 workforce were members of the new FTZ Workers Union (FTZWU), but management refused to recognise the union in the factory. A referendum was held but workers were so intimidated by management that only 4% voted. The union asked for international support and a complaint was filed to the ILO and the Fair Labour Association in the US. Auditors from Nike visited the factory and, through the Clean Clothes Campaign, many people wrote letters to the buyers and to the Government.

On October 16, Jaqalanka met the union and an agreement was reached. The company agreed to recognise the FTZWU as the representative of the workers and to stop all victimisation of union members. The FTZWU promised to stop its international campaign and suspend complaints lodged with the ILO. The General Secretary of the Jaqalanka branch sent the following message 'This is a significant and substantial victory for the brave union members of Jaqalanka and for the workers of Sri Lanka. We thank everyone for their interest and support of this campaign. Your solidarity, support and actions helped make this happen'.

Source: Clean Clothes Campaign (2003:4)

and organises its own programme of activities. These include campaigns directed at particular companies or specific issues, for example a living wage, using leafleting, postcard campaigns and events such as alternative fashion shows. An 'urgent appeal' system is also in place to support workers who are in dispute with management about issues such as the arbitrary dismissal of union organisers or lack of overtime or redundancy pay. Consumers are encouraged to write to employers, retailers and government officials and these interventions are often seen as instrumental in bringing about a resolution to the dispute.

One recent example of the successful use of the CCC's 'urgent appeal' system has been the campaign in support of unionisation at the Jaqalanka factory in Sri Lanka (see Box 3.6).

Through these urgent actions and related campaigns, the CCC has now developed strong links with workers' organisations in Asia and elsewhere. A diverse patchwork of networks now exists between particular European campaigns and organisations in the countries supplying their particular market. For example, the German CCC has developed strong alliances with organisations in Eastern Europe. At the same time, overlapping alliances have been forged with solidarity organisations in other industrialised countries, giving added strength to supportive action, illustrated for example by the victory of the garment workers' union in Lesotho in 2003 (see Box 3.7).

The emergence of 'corporate social responsibility'

Organisations campaigning for workers' rights in the garment industry have mainly targeted well known retailers and branded companies such as Nike, Adidas, Levi Strauss and Gap. This has elicited a change in attitude such that these companies do now accept responsibility for working conditions in their supply chains. One of the first to respond was Levi Strauss, which was exposed in 1992 for employing Chinese prison labour on the island of Saipan. The reaction of Levi Strauss was to introduce a code of conduct on labour standards that was to apply to all the company's supply chains (Shaw and Hale 2002). Other companies followed suit, first in America and then Europe, and by the late 1990s codes of conduct had become the standard response of all major garment retailers to the demands of activists and consumers. Companies drew up a list of labour conditions to be adhered to by their suppliers, relating to issues such as health and safety, working hours, and sometimes the right to organise, that are loosely based on ILO (International

> ## Box 3.7 International links in support of Lesotho garment workers
>
> In January 2001 collaborative research was carried out by the Clean Clothes Campaign and LECAWU, the Lesotho Clothing and Allied Workers' Union, to document working conditions in garment factories. After a press conference to officially release the first research findings, the Lesotho government launched its own investigation into the garment industry, which fully supported the research. Meanwhile, Labour Behind the Label in the UK used the research findings in a campaign to pressure the Gap to take responsibility for working conditions. As a result, the Gap launched its own investigation in one of its contract factories, which resulted in the management recognising LECAWU. Also the Canadian-based Ethical Trading Action Group (ETAG), involving the Maquila Solidarity Network, noted that the Canadian garment retailer Hudson Bay Company sourced from two large garment factories in Lesotho. A campaign was started to pressure Hudson Bay in relation to a wide range of illegal and unfair activities. Nien Hsing, a Taiwanese company, owned one of the factories researched. Nien Hsing had long refused to negotiate with LECAWU. The research reports provided ammunition for a global campaign to put pressure on Nien Hsing, which has been taken up by unions and NGOs in North America and Europe. LECAWU has now been able to negotiate a historic agreement with Nien Hsing, committing management to recognise the union once it organises a majority of the workers.
>
> *Source*: Hale (2004)

Labour Organisation) conventions. The significance of these codes is now a matter of considerable debate amongst academics as well as policy makers (Jenkins et al 2002). As Richard Howitt MEP suggests: 'Suddenly everyone is talking about corporate social responsibility' (Howitt 2002:xiii).

Once companies began adopting codes of conduct, the ground shifted for campaigners and the nature of campaigning activity had to change. Up until then, the focus of campaigns had been on exposing companies and calling on them to take responsibility. Following media exposure, more and more companies acknowledged this responsibility, but claimed that their codes of conduct on labour standards would ensure good practice. Companies could now ward off criticism by publicising their particular code of conduct. The challenge for campaigners, therefore, became to make sure that these codes were more than a public relations exercise and were implemented in practice. In the UK this issue was taken up by a network of NGOs known as the UK Trade Network. Participants at these meetings included WWW and others such as Christian Aid who had begun campaigning for labour rights in the supply chains of UK supermarkets. A subgroup, known as the Monitoring and Verification Working Group, was developed to work on ways of ensuring the implementation and monitoring of company codes. The work of this group led to the establishment of the Ethical Trading Initiative (ETI) in 1998, an organisation managed by representatives of companies, NGOs and trade unions.

The aim of the ETI is to collaborate on pilot monitoring exercises and on other initiatives aimed at ensuring that company codes are properly implemented. A rigorous Base Code on Labour Standards has been established, loosely based on ILO conventions and including the all-important clause on the right to organise. All company members must sign up to this code and undertake activities directed towards its implementation. Most major UK garment retailers have now joined the ETI, in the realisation that it is in their interests to work with NGOs and trade unions rather than be exposed by them in the press. As one of the initiators of the ETI, WWW remains an active member, working, together with the Central America Women's Network and UK-based homeworking organisations, to try to ensure that the voices of women workers are not drowned by corporate interests. Our weakness as small NGOs is offset by our strong links with workers' organisations and the fact that we maintain the potential to expose company practice. This is illustrated by a recent case where WWW brought forward a complaint from the Kenya Women Workers' Organisation about conditions on flower farms supplying UK supermarkets that are ETI members. The subsequent action by the ETI companies was a catalyst in setting up a stakeholder initiative in Kenya, and there are already visible improvements on some of the farms (see Hale and Opondo 2005).

Linking company codes to the demands of women workers

Although WWW remains an active member of the ETI, there are major reservations about any initiative which is based on the implementation of company codes (Hale 2000a). All codes are developed by companies or

Northern-based organisations without any negotiation with the workers they are intended to protect, and, in most cases, they are produced without the knowledge of workers. Once they became widely adopted, WWW therefore decided to take this issue up with women workers' organisations in the network. In April 1998 information about codes of conduct was sent to organisations in Asia, and they were asked whether they wanted to work together on an education and consultation programme on this topic. The outcome was a four-year programme on codes of conduct in Bangladesh, India, Pakistan, Sri Lanka, Philippines, Indonesia and Thailand (Women Working Worldwide 2002a). Funds were also raised for a similar programme implemented through the Central America Women's Network in El Salvador, Guatemala, Nicaragua, Honduras and Costa Rica (Central America Women's Network 1999).

Workers who took part in the codes programme sacrificed their few hours of leisure time to attend educational workshops, which were sometimes held in secret to avoid intimidation. None had previously heard of the codes, though some were working in factories supplying major brand names. Once they understood the purpose and origin of this approach they showed great sophistication in recognising the limitations of company codes, whilst at the same time acknowledging their potential as an organising tool. The general consensus was that codes should not be used as an alternative to collective bargaining and that if workers did not have the right to organise and represent their collective interests the codes were worse than useless. Workers also felt that the implementation of national legislation was more important as a lobbying strategy. However, they recognised the value of codes as mechanisms for trying to prevent companies from moving their sourcing to countries with less rigorous labour legislation (Women Working Worldwide 2002a).

The educational sessions involved in this project also focused on the labour rights covered by codes of conduct, and it became clear that there was a difference between the clauses in codes of conduct and women workers' own priorities. In many cases, educators introduced the idea of codes by getting workers to list what they would put into their own codes on labour standards. This was a similar exercise to that adopted by MEC in drawing up their 'Code of Ethics' for maquila workers, and the results were also remarkably similar. High priority was given to the rights of pregnant women and mothers and to freedom from sexual and physical abuse, issues rarely highlighted in company codes. Most fundamental were concerns about the irregularity of working hours and the lack of employment stability. This was expressed differently depending on workers' employment situations. In the case of workers in the EPZs, one of the key demands was for an end to compulsory overtime, whereas for those in the small workshops of Bombay/Mumbai, the key issue was lack of proper employment status. These workers had no work contracts or even appointment letters and so could not establish themselves as employees (for further information, see Chapter 8, this volume). They recognised that this meant that their ability to claim the other rights incorporated in codes of conduct was extremely limited. Nevertheless, they saw the value of mechanisms that took the issues beyond the patriarchal control of the local employer (Women Working Worldwide 2001:4).

One positive outcome of the education programme on codes was that many workers were introduced to the global dimension of their work for the first time. They were heartened by the realisation that codes had come about because people on the other side of the world were concerned about their working conditions. They also recognised that these people could support them in their demands and so understood the value of international networking. As one Bangladeshi organiser reported: 'Workers became aware that foreign consumers are trying to do something good for them. They have got the feeling that they are not alone. As a result their level of awareness and sense of their rights was raised' (Women Working Worldwide 1999:23). One of the difficulties, however, was that workers' expectations were also raised. Few workers had previously questioned where their products went after leaving the factory, and many became enthusiastic about using brand labels as a way of tracing supply chains and developing links both with other workers and with consumers. Organisers realised that the problem in most cases is that these connections are far from transparent.

Facing up to the challenges of subcontracting and informalisation

The need to understand more about international subcontracting chains was one of the clear outcomes of the WWW codes programme. The participating organisations recognised that no action could be taken in relation to codes unless it was possible to trace the connection between the workplace and the buying company. They also recognised that codes of conduct are unlikely to reach women workers at the end of complex subcontracting chains. It is these workers who are most in need of mechanisms such as codes, and the irony is that they are the least likely to be covered by them. Companies taking action through the ETI, for example, are much more likely to promote the implementation of codes in the first level of the supply chain. Large known suppliers are relatively easy to audit for labour conditions, but regular inspection is almost impossible in the case of small workplaces, and often they are not even covered by national legislation. It became clear that, if codes are to have any impact, one of the biggest challenges must be to address the situation of subcontracted workers.

The organisations in WWW's network have also recognised that subcontracting is increasing and that this poses a major threat to workers' rights in general. Moreover, the increase in subcontracting is seen as part of an overall process of informalisation, which means that, even in large factories, more workers are being employed on a casual basis. This was the key issue addressed at a conference in 2000 in Seoul entitled *Globalisation and Informalisation*, organised alongside the official ASEM (Asia Europe Meeting) by WWW, the Committee for Asian Women and the Korean Women Workers' Association. The conference was attended by representatives of women workers' organisations from fifteen countries in Asia and Europe, and all came with similar accounts of the ways in which global economic integration is accompanied by an informalisation of women's work. For example, following group discussion one group reported:

> The biggest problems faced by women workers are increased job insecurity accompanied by deteriorating pay and conditions. In Indonesia, following the financial crisis, women are now earning only 50 cents and working in bad conditions. In Thailand, job insecurity has become a major issue as many companies are either moving from Thailand or moving to the provinces for cheaper labour. There has also been an increase in the informal sector. In Korea there has been an increase in the number of irregular workers, with 70% of women working in this sector. In Bangladesh there has been an increase in unemployment as factories have moved and there has also been an increase in the informal sector (Da Bindu 2000:6).

It became clear during the Seoul conference that informalisation was the result of cost and risk cutting by companies. It was also clear that these strategies were being used by employers in Europe as well as Asia. As a UK union organiser reported: 'Increasing globalisation has put severe pressure on the UK garment industry. In response some employers run what can only be described as sweatshops. They break many of the UK laws on minimum wage, health and safety protection' (Women Working Worldwide 2000:10). Participants from several countries gave similar accounts of factories being closed and the same work being put out to women outworkers and stories of workers being sacked and re-employed through an agency system. Of course, such measures not only threaten women's rights as workers but also make any attempt to organise and unionise extremely problematic. It was agreed that subcontracting was being used by companies as a strategy not only for reducing costs but also for undermining workers' ability to organise. Although this was not a new strategy, as illustrated by the reaction of IGMC to the actions of women workers in 1989, it was seen as becoming increasingly widespread.

The particular problems of organising in the context of subcontracting were addressed by another conference organised by WWW in the UK in September 2002, entitled *Organising Along International Subcontracting Chains in the Garment Industry* (Women Working Worldwide 2002b). Again it was clear that the increased use of subcontracting was a problem in all the countries represented, which were Thailand, the Philippines, China, Bangladesh, Pakistan, Mexico, Guatemala, Brazil, Romania, Poland, Bulgaria, Palestine and the UK. It was recognised that this increase in subcontracting demanded new approaches to organising that built on the experience of homeworking groups and other community-based organisations. It was agreed that it is important to develop links along the supply chain, so that not only production workers but also office workers, transport and distribution workers, designers and consumers were included. In particular, it was seen as important to build solidarity between unionised workers in factories and those outside. Different ideas were put forward as to how this might be done. Some suggested that: 'The role the trade unions can play in organising workers may be very limited because of the widely dispersed nature of the sector and the high turnover. However what the trade unions can do is to demand in their negotiation with management that certain labour standards should be adopted also when work is contracted out' (Women Working Worldwide 2002b:7). Others suggested unions could become more proactive, and supported the idea of 'community unionism' which involved 'approaching other community, church, anti-poverty, environmental, citizens rights groups, to try to forge alliances with them which give access to new groups of workers' (Women Working Worldwide 2002b:39). At the international level, codes of conduct, social clauses and international framework agreements were all seen as possible mechanisms for collective action.

By the end of this subcontracting conference it became very clear that more information was needed if effective action was to be taken to improve working conditions in the garment industry. Participants all agreed that they needed to know much more about how supply chains work and about the nature of the links between workers inside and outside factories, between local companies and buying companies and between producers and consumers. They also agreed that this knowledge is needed by workers themselves, including subcontracted workers, and that an educational programme could form the basis for more effective organising. Immediately following the conference these proposals were used as the basis for the action research project reported in the rest of this book.

The role of the research in the promotion of workers' rights

The research reported in this publication was thus carried out as an integral part of a wider movement in support of workers' rights in the garment industry. It came about as a result of the expressed need by the participating organisations to know how subcontracting operates in their own locality and to use this knowledge to facilitate their organising and advocacy work. Within this common agenda, each organisation carried out a research programme that reflected its particular place in the workers' movement and the nature of the garment industry in its location (for more information about the research conducted, see Chapter 4, this volume).

Although the nature of the research differed for each participating organisation, they all shared a commitment to building stronger links with workers outside the main units of production. As a result, they have been able to throw new light on the complex nature of outsourcing and the challenges this presents to the labour movement. For example, the layers of subcontracting revealed through the research in Thailand and the Philippines demonstrated the impossibility of organising within the framework of a traditional employer/employee dividing line. A typical strategy was for employers to subcontract out unfinished work to supervisors or workers who would then themselves become agents distributing some of the work to others in the community (for more information, see Chapter 5). For the organisations involved in the research, revealing this hidden hierarchy was not an end in itself. The aim was to use this knowledge to establish new relationships with workers in order to strengthen organising and advocacy work. All were aware of the huge challenges this presents. Such complex subcontracting procedures divide workers from each other and move them a long way from conventional shop floor organising. Whilst remaining fully committed to trade unionism, the participants recognised the limitations of existing models of organisation and the need to promote new organising strategies.

This research extended the reach of the participating organisations not only to local outworkers but also to workers who are part of the same supply chains in other countries and to campaigners in Europe and elsewhere. This was demonstrated by the information uncovered on the Gap. There had been no original intention to focus the research on any particular brand name. The starting point had been the manufacturing companies that were employing workers in the different localities. However, several organisations found they were tracing supply chains up to the same retailer or brand name, notably the Gap. By working from the bottom up, they were able to see that there were common intermediaries in the supply chain, such as regional sourcing offices, and that there were opportunities for solidarity action in support of workers in different countries (see also Chapter 2, this volume). It also became clear that information was being uncovered that would be useful to organisations in the North that were putting pressure on Gap, and indeed to the Global Compliance Department of Gap itself, which acknowledged that the research was beneficial to their internal efforts to bring about changes in the company's buying practices (internal communication, November 2003).

The intention of the research was to serve the needs of the participating organisations but also of all organisations who are campaigning for the rights of garment workers. The CCC and other similar campaigns are focused on raising consumer awareness about the source of their clothing. However, the complexities of subcontracting chains means that it is often very difficult to trace the links between a workplace and a retail outlet. This kind of research can help to fill that gap. The research has less to do with specific information, which tends to change quite rapidly, than with understanding the ways in which subcontracting works and how it is possible to do research to trace the links, for example using the Internet or information on labels. This is also crucial information for those who are in the business of monitoring codes of conduct, since the companies involved have made a commitment to implement these codes all the way down their supply chains. Companies themselves are typically unaware of the extent of these supply chains, as was illustrated in the response of the Gap to this research (internal communication). Yet unless these subcontracting links are known, tools for enforcing corporate responsibility, such as codes of conduct, can have little impact on the very workers who are most in need of support.

Conclusion

The research reported in this publication was carried out as part of a wider movement for workers' rights which has focused on the conditions of women workers in globalised manufacturing industries and the garment industry in particular. The research has thrown up complex issues for all the organisations involved in this movement. However, it is recognised that these issues have to be confronted. Women in subcontracted workplaces are not a few marginalised workers but a substantial and growing labour force that is central to the dynamics of globalised production. Their economic significance gives these workers potential power to challenge these dynamics, but the realisation of this power requires awareness raising and organising work which looks beyond traditional factory floor approaches to union organising. Included in this is the building of stronger links with workers along the same supply chains and between workers and Northern-based initiatives which are supporting their rights.

4

Action Research: Tracing the Threads of Labour in the Global Garment Industry

Jane Wills with Jennifer Hurley

Introduction

As outlined in the previous chapter, a global network of organisations has emerged to support, organise and represent industrial workers in the developing world. If they are to continue in their work of empowering women workers, such organisations need information about the forms of capitalism and the particular global supply chains in which local workers are embedded. As workers' ultimate employers are often located many thousands of miles away at the end of complicated supply chains, it is critical that workers have the means to identify those with power over their working environment and to understand their place in the corporate chain of command. Yet mapping supply chains is extremely difficult and such information is usually confidential, not least because of the risk of political exposure over any local violations of workers' rights. As Naomi Klein (2000:201) puts it in it her book *No Logo*:

> [W]here a previous era of consumer goods corporations displayed their logos on the facades of their factories, many of today's brand-based multinationals now maintain that the location of their production facilities is a 'trade secret', to be guarded at all costs.

Exposing these 'trade secrets'—the sites of subcontracted production—demands that workers and organisers develop sophisticated methods to map corporate supply chains using a combination of published corporate information, the trade press, government records and bottom-up research. In many cases, there is no substitute for the information that can be collated by asking workers about the goods that they make, the sources of the raw materials in their factory or workshop and the destination of finished goods. Using such information, organisers can put together fragments of a bigger picture that stretches far beyond their locale. And, in so doing, organisers can also explore the working conditions of those employed, they can exchange information with workers and try to encourage the development of new organisations.

Such activity takes places within trade unions and workers' organisations everywhere to different degrees, but it is critically important to have the capacity to link such stories and knowledge from one location with those from others. It is no longer adequate to map capital as far as national borders or to the limits of particular firms. In a global economy, workers need to piece together as full a picture as possible of the supply chains of which they are part, identifying possible points of leverage and/or potential grounds for solidarity with others elsewhere. Moreover, workers need to sustain international relationships for the long term, rather than just looking for solidarity at times of crisis, as has often been the case in the past. Thus, as WWW, we have prioritised the development of sustained relationships with a network of organisations, and used this network as a basis for undertaking locally led but internationally co-ordinated action research. In the project reported here this has involved research by 10 organisations working in 9 different countries to map supply chains and their implications for workers in the garment industry.

Unlike academic research that is driven by a particular question, the research reported in this book has been conducted to meet the needs of the 10 local organisations. The projects were designed to collate new information, to reach workers—particularly informal workers sewing clothes for export—and to support education, organising and campaigning activities. Each project had a slightly different emphasis and focus, but it simultaneously fed into the international work. Through its geographical breadth *and* the quality of local relationships that aided the work, the research has produced a wealth of information about the structure of the global garment industry, its impact on workers, and the similarities and differences between various locations.

In this chapter we report further on the nature of the action research undertaken, we explore the mechanics of doing a research project on this scale, we outline some of the key outcomes from the research and, finally, we look at the strengths and weaknesses of the approach.

The Value of Action Research

As its name would suggest, action research aims to bring research and action together. Unlike traditional academic research, where enquiry is driven by a set of questions that need to be answered in order to better understand a particular development, issue or situation, and/or advance knowledge in a particular field, action research is conducted with an eye to both knowledge *and* action: the knowledge is generated because it is needed to assist in changing the world in some way (see Reason and Bradbury 2001; Breitbart 2003; Greenwood and Levin 1998). In the WWW project reported here, such action came via a number of different routes. First, the very process of doing the research empowered those involved as they learnt new skills and increased their self-knowledge and confidence through engaging in the research. Second, the research was used as a means to increase consciousness amongst a particular constituency of garment workers, many of them in the informal economy. The research allowed organisers to share information, make new contacts and get such workers involved in their organisations. Third, the outcomes of the research were used to better understand and then publicise the experience of women workers in the industry and to put pressure on those with the power to act, such as government officials, employers associations and those at the top of supply chains. And fourth, the research can now provide an example for conducting such collaborative projects, illustrating the potential of this type of research to contribute to multiscalar analysis and action over global concerns.

Over the last twenty years, the cultural or postmodern turn in the academy, with its associated emphasis on the importance of power in the production of knowledge, has opened up space for a turn towards action. If power is known to shape the generation of all knowledge, querying any claims to neutral and objective truth through research, there is no reason why producing knowledge for action is less valid than any other kinds of knowledge production (McDowell 1992). If it is conducted to meet rigorous standards, with all necessary attention to the choice and deployment of methods and with careful analysis of the research data, there is no reason why action research should be any less legitimate or insightful than research conducted without explicit political goals. Moreover, the connection to action is particularly helpful as a means to develop the collaborative relationships with research participants on which good qualitative research depends.

As qualitative research relies on face-to-face contact and communication, it is best conducted in the context of trust. If the research is known to be contributing to action, and if research participants themselves have a stake in the action involved, then the quality of the relationships on which the data depends should be greatly improved. As Reason and Bradbury (2001:xxiv) put it, action research can 'show that good knowing rests on collaborative relationships.' Exploring the nature of employment in international supply chains depends on acquiring information from workers on the ground, and they are unlikely to commit to taking part in research that is not politically safe. Workers are fearful of speaking about their conditions of work and, potentially, losing their jobs. In this context, outsiders would have difficulty in finding and then relating to the workers on whom their research would depend. Moreover, such outsiders would have no means to develop long-term relationships with the workers involved and would have no mechanism for improving their conditions of work. In our case, in contrast, action research led by local organisations provided the means to reach workers who would probably not feature in more conventional forms of research, and it also provided a way for these workers to learn more about their situation, and to get more involved in the local support and campaigning organisations that seek to improve their conditions of life.

In practice, action researchers have tended to use a range of research methods, including both quantitative and qualitative techniques, while emphasising the importance of taking a democratic and participative approach to the work. As Reason and Bradbury (2001:xxiv; see also Chambers 1994a, 1994b; Cornwall and Jenkes 1995) put it, action research is: 'a whole family of approaches to inquiry that are participative, grounded in experience and action-oriented.' Rather than having research done to them, action research should allow communities of people to set their own agenda, get involved in the design and implementation of the research, and own the information that is produced. In their account of the participative research methods that are generally adopted when undertaking action research, Cornwall and Jenkes (1995:1668, emphasis added) argue that:

The most single striking difference between participative and conventional methodologies... lies less in the theories which inform these methodological frameworks or even in the methods they use but *who* defines the research problems and *who* generates, analyses, represents, owns and acts on the information that is sought.

In the WWW research reported here, the projects were participative in a number of ways. First, the members of each organisation were able to design their own research project, to meet their own needs. Although as WWW we secured funding for the work and provided support and general information, each local organisation was left to define its own particular research requirements and to design an appropriate research strategy in order to fulfil these needs. Second, the women employed by the local organisations and some worker-researchers benefited from being exposed to the research process, learning new skills and taking part in international-level exchange. And third, each organisation conducted research that was never simply 'top-down.' In each case, workers were interviewed in order to collect information, but these encounters were also opportunities to share knowledge about workers' rights, to provide information and support, and to encourage collective organisation. In a number of cases, workers were also involved in workshops to discuss the research, its implications and the possibilities it posed for action. As much as possible therefore, these 10 different research projects conducted in 9 different countries were organised as 'bottom-up,' participative, action research encounters.

This approach to information collection is widely adopted by non-academic organisations and activists, even if it is not labelled action research. When a political organisation needs to find out information in order to conduct a campaign, for example, it will routinely use its own staff and activists to go out and talk to people, conduct surveys and assemble the information needed to plan and implement the campaign. In some cases, groups might choose to work with sympathetic academics and/or students to do this work, and there are a number of universities that have teams of researchers who work in this way. In the UK, examples include the Centre for Public Services at Sheffield Hallam, the Centre for Public Policy Research at University College London and research work conducted by the East London Communities Organisation (TELCO) and Queen Mary, University of London (see Wills 2001b, 2003b, 2004; Wills et al 2002; and for examples from the USA, see Nyden et al 1997).

Workers' organisations are particularly well placed to take part in action research projects. They generally need information to inform negotiating and organising strategies and have a wealth of human resources and contacts on which they can draw in this work. A recent example of internationally co-ordinated action research in this vein is that led by Homenet, an international network of organisations that support and organise home-based workers. Their research has been designed to map home-based work both horizontally (in place) and vertically (in its connections to local, national and international trade) in a number of countries in Latin America and Europe. This project has deployed action research techniques developed by local NGOs, unions and academics to help establish new organisations of home-based workers on the ground. As the project's research advisor, Ruth Pearson (1994:143) puts it:

> The project's methodology is to collect information about home-based work in a particular location, to work with home-based workers to analyse their situation and to identify problems and priorities for action, and to devise plans for advancing strategies for change.

In a similar approach to that adopted by WWW, Homenet has co-ordinated research activities in a number of different countries with the aim of collating new information *and* strengthening local organisations that are able to support, represent and organise homeworkers in future.

Transnational Action Research Across Ten Organisations in Asia and Europe

The organisations involved in WWW's garment subcontracting chains research work are listed in Table 4.1. Some are locally based women workers' membership and/or advocacy organisations such as Karmojibi Nari in Bangladesh, Friends of Women in Thailand, Innabuyog Metro-Baguio in the Philippines and the Working Women's Organisation in Pakistan; another is a Christian based NGO (the Hong Kong Christian Industrial Council); others are locally based research groups focused on workers' rights (the Union Research Group in India and the Philippine Resource Centre); and the remainder are labour-oriented organisations with their own local and international links (the Transnational Information Exchange in Sri Lanka and the Bulgarian-European Partnership Association). All have medium- to long-term relationships with WWW and all have a commitment to supporting workers towards their own self-organisation. In most cases, garment workers are involved in the day-to-day work of the organisation, and in some the leading activists are, or used to be, directly employed in garment production.

As part of the international project to better understand subcontracting and supply chains in the garment industry, each organisation developed its own research objectives, as listed in Table 4.1. As is evident, the research done by each organisation varied considerably. While the Philippines Resource Centre, for example, wanted to map the chains of the top 20 multinational garment companies in the country, the Sri Lankan organisation chose to focus on homeworking and the links between the production of brand-name goods and those working at home. These differences reflected gaps in existing knowledge. In Sri Lanka, for example, the local organisers knew quite a lot about conditions in the large factories for export, but very little about the extent and conditions of homeworkers, whereas in the Philippines they wanted to chart the links of the biggest garment producers. As producing knowledge and supporting action were key aims of the research, WWW had no wish to impose a tight research agenda or model on each organisation. If they were to act, each local group had to determine its own set of research questions and devise suitable methods with a view to achieving its goals.

Although WWW set no overall research framework and saw it as unnecessary to train partners, each local organisation was not left on its own. Staff from WWW in the UK played a critical role in supporting the research, providing administration to the network and providing the means to share ideas, experiences and research findings. Briefing papers were provided about action research, supply chains and research techniques. There was no obligation to pay attention to these, unless they were seen as useful. Some organisations, such as the partners in Bulgaria and India, included members with academic backgrounds who developed their own theoretical perspectives on the material provided. Others, such as the groups in Thailand and the Philippines, grounded their research more in a worker-based approach to political action. WWW also helped to collate information about the particular multinational companies that were identified on the ground and encouraged the use of Internet and business information libraries to share information about a company's strategy, its corporate social responsibility policies and its geographical reach.

In addition, WWW liaised with local partner organisations to arrange two project workshops and two regional research meetings. In this way WWW allowed each local research project to develop separately while being simultaneously linked to the others in ways that increased the insight and impact of the research as a whole. The two regional meetings were located in South Asia and South East Asia and were held six months into the first year of the project. The purpose was for the project co-ordinator and one researcher from each group to meet and discuss

Table 4.1   The organisations involved in the research and their research objectives

Organisation	Main activities of the organisation	Research objectives
Bulgarian-European Partnership Association, Bulgaria	Research Bulgarian garment industry. Increase awareness of the problems faced by workers.	Assess pay and conditions in industry. Map subcontracting chains of several international companies, including homeworkers where possible.
Hong Kong Christian Industrial Council, Hong Kong	Support the rights of workers, particularly the marginalised and poor. Run campaigns, worker education, training and advocacy.	Map subcontracting chain of Hong Kong-owned companies in mainland China. Document the working conditions of workers, especially home/informal sector workers and their opinions on improvements.
Friends of Women, Thailand	Support women workers in manufacturing industries. Promote the rights of women facing violence and discrimination. Provide counselling, legal assistance, training and support for women workers' organising.	Learn how the target factories implement subcontracting. Collect information on where power is located in the chain. Understand how producers and agents operate.
Karmojibi Nari, Bangladesh	Strengthen women's participation in trade unions and promote equality in the workplace. Build solidarity among women from different occupations.	Compare quota and non-quota companies. Map the overall trend of the industry. Document worker issues.
Philippine Resource Centre, Philippines	Support workers in both the formal and informal sector through research. Inform groups in Britain about issues of political, social, economic and cultural importance to the Philippines.	Map subcontracting chains of top twenty foreign TNCs. Build links between workers in factories and sweatshops and homeworkers.

Table 4.1 (*continued*)

Organisation	Main activities of the organisation	Research objectives
Transnational Information Exchange Asia, Sri Lanka	Encourage the formation of democratic unions and more leadership by women, and facilitate broader social coalitions. Run training and education programmes for workers.	Map the subcontracting chain of Sri Lankan suppliers to home-based workers. Explore the relationship between brand label producers and the informal sector.
Union Research Group, India	Support the development of genuine trade unionism. Focus on women workers. Organise education and discussion sessions for worker activists.	Map the structure of the Indian garment industry and export production links in the subcontracting chain. Document employment conditions of workers in informal sector. Research legislation.
Working Women's Organisation, Pakistan	Strengthen the capacity of working women to defend their rights. Organise education centres, training programmes and undertake advocacy work.	Trace the role of contractors, middlemen and companies. Research conditions for women workers (factory and home-based) in the chain.
Innabuyog Metro-Baguio, Philippines	Support the organising capacity of women workers. Carry out education and leadership training for women's organisations.	Trace supply chains from factories in the Baguio City Economic Processing Zone. Document conditions of knitters and weavers who work from home. Explore ways for contract and permanent workers to organise together.
Women Working Worldwide, UK	Support the rights of women workers in a globalised economy. Support an international network of women workers' organisations that promotes solidarity. Exchange information, organise international meetings and represent the demands of women workers.	*Understand the relationship between the decline of the industry and the nature of subcontracting in the UK. Explore the key challenges facing workers in the industry.

* In addition to providing overall co-ordination and research support to the organisations involved, WWW had its own research project, looking at the situation in the UK.

their research. Participants exchanged information on the supply chains that they were researching, the methods that they were using and the themes that were beginning to emerge from the workers' interviews they had completed.

These regional research meetings were then followed up by two international meetings involving all project participants. The purpose of the first, held in Bangkok, Thailand, at the end of the research phase in February 2002, was for each group to share its findings and discuss ways of disseminating the information at the local, national and international level. Discussions focused on supply chain structures and the key issues that emerged from the workers' interviews. The meeting then looked at practical strategies for disseminating the research findings, developing educational materials and building solidarity between workers within and between countries. As the representative from WWO (Pakistan) said: 'employers in Pakistan and Bangladesh are co-operating, so the workers need to learn to co-operate together too, and move beyond the fears and tensions that have kept us apart.' The second international meeting was held in Manila, The Philippines, in October 2003, and focused on the development of educational materials based on the research. Participants came together with a worker educationalist commissioned by WWW to discuss the piloting of draft education materials. The meeting also discussed the outcomes of the research in each country.

Thus the research was co-ordinated internationally, but decision making and activity were decentralised to the local level to allow the organisations to complete research work that would be most useful to them. The work of the Transnational Information Exchange (TIE Asia) in Sri Lanka provides a good example of this method of working, illustrating the ways in which the research objectives, strategy and methods were locally determined (for a summary of the research conducted in the Philippines, see Box 4.1; and in India, see Box 4.2). TIE decided to work in partnership with two other local organisations that were directly supporting garment workers, Da Bindu and the Women's Centre, and the three organisations met to discuss the research. Each wanted to know more about the extent to which informal employment and homework contributed to the production of garments for export, in order to augment anecdotal evidence that such practices were already widespread. Kelly Dent from TIE Asia reported that they produced some clear objectives for the research, and these were to:

Box 4.1   Research conducted by the Women Workers' Project (Innabuyog-Metro) Baguio City, Philippines

The research objectives of the Women Workers' Project (WWP) grew from earlier work into codes of conduct and the problems that it faced organising garment workers in Baguio City. In this research project, its research objectives were:

1. to trace the subcontracting chains from two factories in the Baguio City Economic Processing Zone
2. to describe the situation of workers along these chains and to contrast this with the situation of workers in the traditional hand weaving industry
3. to use this information to create education and organising strategies among both groups of workers

The research was conducted by three women who were organisers with WWP. As they were not trained in research, they sought assistance from other similar groups to design and conduct the research. The process began with a number of education workshops for the research team. These workshops were carried out over one month and involved intensive discussions about globalisation, its impact on workers, the local situation facing trade unions and the urban poor, basic workers' rights and the contractualisation of workers. The aim was to ensure that the research team thoroughly understood the wider context that had contributed to increased subcontracting, as well as its impact and implications.

WWP viewed the research process as a two-way conversation, where the interviewers would be able to provide the workers with as much feedback, information and assistance as possible. The research took the form of personal visits to interview key informants and conduct focus group discussions. The workers interviewed were generally members of a women's association or an urban poor association. New contacts were very hesitant to be interviewed and it took a long time to gain their trust. Most interviews were scheduled around the women's work, so they were conducted late at night or on Sundays. Personal interviews lasted 2-3 hours and focus group discussions lasted 4-6 hours. Both methods followed a 'question and answer' structure that often resulted in animated discussions.

(*Continued*)

Box 4.1 (*Continued*)

One of the biggest challenges faced by WWP was analysing the volume of data that it collected. It found that this took longer than collecting the data. It also discovered that it had gained a lot of data that was relevant for its on-going work, but not relevant to the research project. Consequently, dissecting the data and documenting it took considerably longer than WWP had anticipated.

To illustrate the findings, WWP presented case studies of different workers' situations. One of the most interesting findings, which is a product of the bottom-up approach, was the blurring of the distinction between employee and employer. The case study outlined below provides an excellent example of this emerging phenomenon. WWP's research also provided exact piece rates that provide insight into the level of underpayment in supply chains, even those headed by famous brands.

Case study: Geri a homeworker supplying an international brand

Geri proudly explained to us how she was able to finish an average of four bundles a day, how her beautiful children help her work on the bundles even though they do not get paid, how important it is for her to earn an additional P88.00 or US$1.64 per day and how other women in the community do the same.

About 80–100 sewers in this one community alone work for the same subcontractor. They have to do the tedious job of sewing tassel ends for mufflers and shawls. The other women are grateful to the subcontractor who has given them this work and pay. But Geri is not. For she knows the subcontractor will get paid P74.40 or US$1.40 per bundle of tassels sewn.

The subcontractor gives the women workers P22.00 or US$0.41 for each bundle sewn, allowing the women to earn a meagre US$1.64 per day, while the subcontractor (agent) earns for herself, from their sweat, US$3.96 per day per sewer (assuming that each finishes an average of four bundles per day). If this subcontractor has a minimum of 80 sewers, she then earns approximately US$316.80 per day.

US$1.40 / bundle Price paid by the company to the subcontractor (agent)

Box 4.1 (*Continued*)

US$0.41 / bundle Price paid by the subcontractor (agent) to the homeworker
US$0.99 / bundle Profit for the subcontractor (agent)
In eight hours:
4 bundles @ US$0.41 per bundle = US$1.64 per day for the home-based worker
4 bundles @ US$0.99 per bundle = US$3.96 per day for the subcontractor/agent
If the subcontractor (agent) has 80 home-based workers and gains US$3.96 profit from each home-based worker each day:
US$3.96 × 80 = US$316.80 per day for the subcontractor (agent)

Box 4.2   The research conducted by the Union Research Group, India

The two researchers in the URG had already built a relationship with most of the workers that they interviewed through the previous project on codes of conduct. The advantage of this was that the workers trusted the researchers and did not fear that they would lose their jobs if they were interviewed. They found that the limitation of this relationship was that, given the extreme tensions in the garment industry, it was impossible for the researchers to interview employers, as the workers would have felt suspicious and uneasy. In retrospect, the URG researchers felt that if they were doing the research again, they would hire an external researcher to approach the employers.

The aim of the research was to trace the three supply chains that employed the women with whom the URG was working. As part of their research, the team discovered that one of the companies was part of a consortium that was being investigated for anti-union activities in the US and another company was supplying a Dutch retail chain that was being audited by the Fair Wear Foundation.

The URG used a threefold research approach. They carried out interviews with factory and office-based workers, they got information from the two organisations auditing two of the companies and they did Internet-based research. Despite this multifaceted approach, they still found there were large holes in their research that were difficult to fill. For example, they obtained a comprehensive list of buyers, but, because they could not find out how much each buyer

(*Continued*)

Box 4.2   (Continued)

ordered, they were unable to identify the balance of power in these relationships.

The researchers used collective interviews and group discussions to carry out their research. They felt that these methods were valuable in that they provided a collective check on the information and they shared the 'pool of information' among the workers, so that they all learnt about the complexity of the chains and they were able to help each other piece the information together. The workers were paid for the time they took off in order to participate.

Another challenge that the researchers faced was that many of the homeworkers expressed a desire to organise. Because the researchers were providing them with information about their rights and how to obtain them, the workers assumed that they would be willing to help them organise, even though that is not the brief of the URG. To overcome this, they put the workers in contact with someone who was willing to help them organise, but the researchers did wonder how willing the workers would have been to do this if they had seen the researchers as 'as just . . . gathering information , or even as providing education which was of no practical relevance to their employment conditions.'

This experience shows the complexity of relationships between researchers and workers in action-based research, even when trust had been developed. In this case, research that can help meet more long-term needs, for example by raising consciousness through educational work, may not serve the short-term needs and interests of the workers. In some situations, there is a need to balance different needs and desires and to find ways of, if not accommodating conflicting needs, then at least hearing those needs and attempting to point out alternative sources of assistance.

- Determine the extent of informal employment in the garment sector in Sri Lanka
- Discover the issues faced by workers and their suggested solutions
- Make contact with more informal workers and give them information about their rights
- Inform organising strategies (including campaigns) for workers in both the formal and informal sectors

- Create links between workers in formal and informal employment
- Provide useful information to unions
- Help develop the programmes and activities of TIE Asia, Da Bindu and the Women's Centre

In most cases, the organisations involved did not have extensive experience of conducting research, and the local organisers decided to use the project as a means of increasing their skills. In the case of TIE Asia, the project provided an opportunity for training a team of workers from each of the three organisations to undertake the research. They assembled a group of researchers with complementary skills, including those with some prior experience of data collection, interviewing, engaging with workers and report writing. Once twelve researchers had been identified, training was provided to introduce the team to ideas about globalisation, supply chains and subcontracting, before going on to look at the practice of research itself. Interviewing formed the backbone of the research work, and researchers spoke to workers about their conditions of employment, the garments they produced and what they knew about the origin and eventual destination of these products. Finding research participants was not easy, not least because workers employed in the informal sector sewing for export are often hidden, working in small workshops and homes. Researchers had to try and catch workers leaving workshops and as they travelled to work, but on occasion this led to confrontations with managers and factory owners. Such circumstances also made it difficult for the workers, and sometimes it was difficult to know if workers felt relaxed enough to tell the researchers the truth.

In the end, the researchers conducted 207 questionnaire-based interviews with workers (of which 185 were included in the final research) and found participants by snowballing: asking each worker to pass on the names and contact details of any friends and relations they knew in the sector. As the interviews were conducted by staff and activists from Da Bindu and the Women's Centre, most of whom had personal experience of working in the industry and many of whom were also migrants to the area, the researchers were generally able to relate to, and empathise with, the workers they met. They could also use the encounters as an opportunity for dialogue, and workers were encouraged to ask questions about their work and their rights. In addition, the interviews provided an opportunity for workers to get to know representatives from Da Bindu and the Women's Centre, organisations that could provide advocacy, support and a route to activism should they need or chose it in future.

All the organisations involved in the international research network developed their own capacity to undertake the research, some buying in outside support and others, like the Sri Lankan organisations, empowering their own staff. Most relied on interviewing workers to find out more about subcontracting and its impact on employment in their locality, and many started with the links between larger factories and the global economy. By looking at one factory, researchers could find out about the connections up the chain, to the main buyers, retailers and merchandisers, and down the chain to where further subcontracting and homeworking took place. The researchers were able to find out about the terms and conditions of workers at each stage of the chain and start to have some dialogue with the workers making the goods. As is evident from Table 4.1, a number of the organisations saw the research as an opportunity to improve their contact with, and knowledge of, those employed in smaller workshops and those working at home. All the organisations had previously found it difficult to contact workers in the informal economy, and, in a number of cases, this research was seen as a way to overcome these problems of access.

Once contact was made, however, workers still needed to be reassured about the nature of the research before they agreed to take part. The Working Women's Organisation (WWO) in Pakistan, for example, found that many workers had real fears about taking part. Some would express doubts about the research, they might refuse to answer sensitive questions and/or express their fears about losing their jobs if they were identified through the research. To help overcome such fears, the WWO made contact with workers through trusted intermediaries such as partner NGOs, community groups, friends and trade unions. They also conducted the interviews in places where workers felt safe, such as their homes and community centres. This also provided the opportunity for researchers to meet other family members and offer further reassurance to them. The researchers shared as much information as possible with workers, both about the activities of WWO and about the research project itself.

Education and Action: Outcomes from the Research

In addition to the generation of knowledge, the success of action research is judged on the basis of the impact it has on those involved and the world around them. As Bradbury and Reason (2001:448) put it: 'A mark of quality in an action research project is that people will get energised and empowered by being involved, through which they may develop newly useful, reflexive insights as a result of a growing critical consciousness.' In our case, the projects led to enhanced research capacity in each organisation; to an improved skill-base for those individuals who conducted the work; to the production of new material to inform political action; to additional educational work amongst women garment workers; and to the development of new organising initiatives. It also gave us the opportunity to improve our understanding of the global garment industry by piecing together the information gathered by each local project. This overview was discussed in project workshops and presented in reports, and much of the information collected is set forth in the rest of this book. Here, however, we explore the ways in which the research assisted each organisation to conduct education and action at the local scale.

As outlined in Table 4.2, each research project generated a diverse set of outcomes, but all focused on disseminating research reports, holding educational workshops with women workers (sometimes in partnership

Table 4.2 The main activities following the research

Organisation	Main activities following the research
Bulgarian-European Partnership Association, Bulgaria	Dissemination of research findings and presentation at national and international forums. Educational seminars for workers held in partnership with a trade union. Helping to develop new plans for organising.
Hong Kong Christian Industrial Council, Hong Kong	Setting up of a workers' centre from which education programmes have been developed on globalisation and supply chains. Distribution and use of research by other organisations supporting workers in China and elsewhere.
Friends of Women, Thailand	Dissemination of research report to workers, unions, NGOs, employer associations, government and media to further debate. Development of a new programme to identify and instruct trainers about subcontracting and codes of conduct. Production of a handbook for trainers and an educational game.
Karmojibi Nari, Bangladesh	Wide dissemination of research reports, including a short report for workers. Linking research on subcontracting with work on the WTO and MFA. Production of associated booklets, posters, and calendar. Training sessions involving hundreds of workers. Improved links with trade unions and new international connections for campaigning.

Table 4.2 (*Continued*)

Organisation	Main activities following the research
Philippine Resource Centre, Philippines	Production of research report and new educational materials. Research presented to workers, unions and other NGOs, and integrated into trade union training programmes. Development of specific educational sessions for workers.
Transnational Information Exchange Asia, Sri Lanka	Wide dissemination of research report, including NGOs, trade unions, government bodies and employers. Press conference and workshop to further disseminate the research and formulate demands relating to informal workers. Production of a booklet and poster outlining the rights of informal workers taken to boarding houses and centres of informal production. Educational programmes for formal and informal workers.
Union Research Group, India	Distribution of the research report to unions, NGOs and researchers. Workshops held in Bombay/ Mumbai, Chennai and Delhi for workers, union activists and NGOs. Launch of a new campaign to set up a Garment Industry Tripartite Board to register workers and thus secure their rights.
Working Women's Organisation, Pakistan	Wide dissemination of reports and workshops for trade unions and NGOs. New educational materials used in workshops and training sessions and included in newsletters.
Innabuyog Metro-Baguio, Philippines	Dissemination of research findings. Research integrated into training modules for women workers. Media exposure of the situation facing women in subcontracting chains.
Women Working Worldwide, UK	Production and dissemination of the following materials: 1. Full research report: Garment industry subcontracting and workers' rights. 2. Report about conditions in the UK. 3. Research handbook: Action research on garment industry supply chains 4. Booklet on Gap 5. Education pack. Further dissemination done through the publication of this volume as part of a follow-up project funded by the EC to work with the labour movement and others in the EU on workers' rights in supply chains.

with local trade unions) and planning new strategies for organising garment workers in defence of their rights.

Most of the participating organisations held educational sessions with women garment workers to help them explore their role in the global industry, what this meant for their lives and work, and the different actions they can take to improve their situation. As the representative from the Working Women's Organisation in Pakistan put it: 'Through this project, workers are learning about their rights, the TNCs' subcontracting systems and the importance of collective struggle for getting their rights.' Although in some countries, like Bangladesh, the local organisation already had a structure of local branches where women could meet and discuss their situation, in other places, new networks had to be established, often for the very first time. In Bulgaria, for example, such sessions were found to be such a useful forum for workers to exchange their own experiences and ideas that it is planned to continue these meetings in future.

WWW worked with participant organisations on the development of a pilot education pack to help them make the most of the research. Focused on workers, the pack was designed to enable them to understand where they fit into global supply chains, where power lies, and possible ways of defending and improving their rights. In practice, each organisation then developed this pack to meet their own needs as educators, and to engage with the women garment workers in their locality. In Thailand, Friends of Women created a new game called 'Your World' in which they explained globalisation, supply chains and labour-related issues through play. In Pakistan, the Working Women's Organisation used videos, and elsewhere new diagramming and mapping techniques have been used. In a number of cases, the material also helped to inform the training courses provided by trade unions in the sector.

Many of the organisations involved found that the research helped them to emphasise the connections between so-called formal and informal workers in the industry. As a representative from the Working Women's Organisation in Pakistan put it:

> The research findings made a link between the formal and informal sectors. One vital link between both sectors is that they are working for TNCs and almost facing the same problems. This link would help us to organise workers and convince trade unions to emphasise the need to organise informal sector workers.

Research like that conducted in Pakistan has allowed NGOs to highlight the contribution that informal workers make to the global economy, to illuminate the links that exist between them and the formal economy, and to convince people of the need to organise these workers. Trade unions have often viewed such workers as unorganisable, and the research has been used to put pressure on trade unions to shift their approach. In Bulgaria, for example, the NGO involved in this research has been working in partnership with a local trade union to try and develop a more sophisticated organising strategy. It is hoped that by understanding the supply chain, local trade unions can identify the existence of codes of conduct and use this information in organising campaigns. Deeper understanding might also mean that unions come to broaden their scope beyond traditional factory workers to include all those employed along particular subcontracted supply chains. By working together, unions and NGOs can share their expertise in mapping and organising workers employed in such chains (for examples of unions and NGOS working together, see the special issue of *Development in Practice*, 1994; and for examples of organising along supply chains in the North American auto industry, see Holmes 2004). The link to WWW also makes it possible to build links to campaigners in key markets, where most consumers and the HQs of most large corporations are located.

In some cases, the research had a significant impact on the work of the organisations carrying out the research. For example, the Union Research Group in India acknowledged that 'This research has had quite a profound effect on our organisation, in the sense that URG previously only worked with unionised workers in the formal sector... Working with non-unionised informal workers throws up the whole issue of organising. Information and even education of workers makes very little sense unless the knowledge can be used to improve their conditions.' In a number of other cases, the research was used as a mechanism for beginning to build such connections with a view to organising in future. In Sri Lanka, for example, the informal-sector workers who were interviewed were asked about how they wanted to follow up the research; over half said that they wanted to learn more about their rights, and many said they would like to be involved in an organisation that would help them with their work problems in future. As a consequence, there is a now a commitment to extending the work of the participant organisations to help them in meeting these needs.

In this vein, the two organisations involved in the Sri Lankan research prioritised the distribution of information about informal workers' rights. Organisers visited areas, villages and boarding houses where informal workers live in order to develop relationships with these workers. The research allowed organisers to get into relationships with informal workers and invite them to workshops and events held at their centres. As Kelly Dent, organiser with TIE Asia put it:

> The project has been extremely valuable in learning about the issues facing a largely invisible section of mostly women workers in the garment indus-try employed informally. Understanding how the law in Sri Lanka relates to informal employment has been important. This increased understand-ing has led TIE Asia, the Women's Centre and Da Bindu to alter their work and to include women employed informally in their programmes, cam-paigns and activities. The research has been invaluable in including the issues of workers informally employed into broad-based campaigns and raising their issues generally within the trade union movement and labour organisations in Sri Lanka... Most importantly, it has led to resources being developed for workers in informal employment, and training pro-grammes have been developed to raise awareness of their rights.

In India, the research highlighted the problems facing those in the informal sector as, without registration as workers, they had no means of accessing their rights. India has a particular problem here as the vast majority of garment workers in the export sector are found in the informal sector, which is not the case in other locations such as Sri Lanka and Bulgaria. In India, the Union Research Group thus decided to launch a new campaign, drawing on the research, to demand that a Garment Industry Tripartite Board be established to register informal workers. As their representative explained:

> We plan to use the research materials to build up a campaign for employ-ment regulation in the garment industry so that the vast majority of workers, who today are informal, unregistered and have no rights, will ultimately be registered as workers and have their rights recognised and protected. This is not going to be easy in the current climate, where even the existing labour laws are being contested by employers and diluted by the government, but at least workers can be made aware of what rights they should have and provided with some support if they wish to fight for them.

The Union Research Group has used the research to hold meetings (in Bombay/Mumbai, Chennai and Delhi) with workers, other NGOs, trade unions and labour activists to initiate this campaign. Moreover, through the link with WWW and the other research projects, it has been possible to integrate this initiative with developments in other locations, as the representative from the Union Research Group continued (for further information, see Chapter 8, this volume):

> Information from participants in other countries that are part of the project has been crucial in order to get a perspective on what is going on elsewhere, and also what it is possible to achieve. Although the campaign we are planning is in India, to some extent its success depends on similar efforts being made in other countries, in international bodies like the ILO and so forth... We intend to pursue the campaign coming out of this project as a kind of test case to see whether this method of formalising employment can be used more widely to help informal workers to get their rights.

Many valued the international-level network that developed through working with WWW on this project, not least because it allowed local activists to understand their situation in its wider context. As a representative from the Philippines Resource Centre put it: 'The international scope of the project provided us with a broad perspective and understanding of how sub-contracting chains operate globally... and provided precious start-up knowledge of the general features and mechanisms driving the whole global chain.' Rather than each organisation having to start from scratch, the network pooled resources and allowed activists to share knowledge, making it easier to work with confidence on the issues faced on the ground.

Outputs from the project include new materials for activists seeking to do similar work. A manual entitled *Action Research on Garment Industry Supply Chains: Some Guidelines for Activists,* was produced, drawing on lessons from the research to share with others. This includes information about how to plan research, how to find corporate information on the Internet, how to organise interviews, and how to present information, as well as the addresses of useful websites. Copies of questionnaires and interview schedules are also included. In addition, the research reports from each of the local organisations have been compiled into one document and a short report has also been produced on the Gap (see Hurley 2003). This material has helped WWW to better understand trends in the industry and has assisted in the intervention WWW is able to make to improving employment standards through its work with the Clean Clothes Campaign and the Ethical Trading Initiative in the UK. The research has also enabled WWW to enter dialogue with the Gap, not least because the company was not fully aware of the extent of subcontracting in its own supply chain. Following the publication of WWW's pamphlet on the Gap's supply chain, the company has started to discuss ways of improving its systems in future. Furthermore, as WWW, we are using the information to support educational work with unions and workers' education programmes in the UK and Europe, highlighting the implications of globalisation for workers.

The Challenges of Conducting Internationally Networked Action Research

Co-ordinating research on this scale, involving a network of grass-roots organisations, is no easy task. There can be no doubt that a research project that spans two continents will present challenges and opportunities that extend far beyond those faced in more conventional types of research. Moreover, this is so much more the case when those involved in the research are not professional researchers, but activists with on-the-ground experience of the issues but little, if any, practical knowledge about doing research. However, because the research evolved from previous experience in the WWW network, the shared history of working together on common issues, engaging in discussions across geographical boundaries and having a good understanding of each other, meant that the project foundations were built on long-standing relations of co-operation and trust. This trust helped to overcome the challenges of working across so many different organisations in so many locations. In addition, the fact that each organisation was able to determine its own research plan and to integrate the work into ongoing activities, made it easier to manage research on this scale.

The devolution of decision making also helped to improve the quality of the research material produced. Using trusted local organisations to gather information about the global garment industry helped to reach workers, particularly informal-sector workers and homeworkers, who are often so difficult to locate. It also made it easier to overcome any fears or misgivings that such workers had about the research. By forging relationships that were to be sustained over time, researchers were able to develop two-way exchanges with workers and unearth information that would otherwise be very difficult to obtain. Examples include findings about the extent of workplace oppression, the degree to which work is subcontracted, the impact of working life on family and home, and the ways in which fellow line workers also act as agents for the distribution of homework in some locations (material that is further elaborated in Chapter 5, this volume). If, as action research advocates suggest, good knowing rests on collaborative relationships, we argue that the research relationships developed in this project facilitated the production of good-quality knowledge about the global garment industry and its impact on workers.

Because the research grew out of the needs of the organisations, it was also possible for WWW to manage things with a very light touch. Even though feedback was minimal at times, it was known that the local organisations were committed to the project, the research and its outcomes. The shared history of having worked together in the past made it easier to develop this 'hands-off' approach to managing the project, and this was more congruent with the political agenda of the research, allowing the organisations to make their own decisions about what worked best for them. As the work got underway, questions would be raised with WWW via email when necessary, but most of the co-ordination, feedback and exchange took place at the international and regional meetings.

Overcoming geographical distance and communication problems was a particular challenge in doing the research. Although some members of the organisations knew each other from previous work, communication was necessary to build links between the groups and to generate discussion about the research. This was done through the face-to-face meetings, but language differences sometimes prevented easy exchange. Although many of the participants spoke English, some relied on interpreters, and it was sometimes difficult to involve them fully in informal discussions. Added to this, the aim of building strong international links was threatened by the requirement that activists had visas to travel. Some researchers found that they were barred from travel at the last minute. As President Bush was visiting the Philippines shortly after the second international meeting in October 2003, those organisations based in Islamic countries were unable to send representatives.

Another challenge lay in the different research objectives of the various organisations, as this made the analysis and synthesis of the research reports problematic. In turn, however, this diversity was central to the depth and richness of the research. The differing research aims and outcomes all revealed partial maps that combined to create a more complex and nuanced picture of subcontracting in the global garment industry.

The research had a significant effect on all who were engaged in the process. Action research is most transformative when it helps to meet the real needs of local people and their organisations. It works best when the research and the political work being conducted by an organisation are mutually reinforcing. By training workers in factories in Thailand to become researchers, for example, Friends of Women (FOW) ensured that the interaction between research and action are set to continue long beyond the life of the project. Likewise, by extending the concept of a research interview beyond a researcher asking questions to encompass a two-way conversation, WWP in the Philippines developed a growing awareness that research is about learning and sharing. They found interviews to be an empowering political action by developing a two-way exchange. Action research effectively creates learning communities, because the relationship between those doing the research and the 'subjects' of the research is not focused on the acquisition of knowledge by the former, but on sharing experience and developing forms of engagement that deal directly with the challenges faced on the ground. As a result, the process and outcomes of action research are more deeply embedded in the communities involved. Both the process and the outcomes of the research are infected with a political engagement which can be both subtle—such as changes in attitude or affect—or more dramatic, where new ideas, campaigns and action result from the research process and findings.

Conclusions

This research demonstrates the way in which action research can be co-ordinated internationally while also allowing local organisations to meet their own aims and objectives. It has highlighted the value of empowering local organisations to develop their own research so that they can conduct work that is appropriate to their needs. In our case, this approach has produced a rich account of trends in the global economy, but it has also helped to empower those organisations that are best able to support and organise export-oriented workers on the ground.

All too often, research is undertaken to look at developments in, and connections between, one or two geographical locations, but in a globally integrated economy we need to construct knowledge on the basis of greater geographical breadth. Moreover, we need the tools to do this without sacrificing the depth of local knowledge and the personal connections needed to complete good qualitative research. Devolving research to local organisations allows researchers to produce in-depth accounts of the world, drawing on their own experiences and relationships, while also feeding into the production of knowledge at a much broader scale. In our case, local researchers have also been able to put their new knowledge into action and to develop education and organising strategies amongst workers. This model of international networked action research—which has the geographical breadth *and* depth that can come from working in collaboration with local organisations—could be an increasingly valuable tool in the battle to understand and challenge the impact of globalisation. Moreover, deploying action research can produce knowledge of greater clarity, intensity and relevance than that led from above.

5

Unravelling the Web: Supply Chains and Workers' Lives in the Garment Industry

Jennifer Hurley

Introduction

The garment industry is composed of unseen, intricate supply chains that run like invisible webs around the globe. As outlined in previous chapters, these chains have evolved in the drive to achieve low-cost, low-risk and flexible production in an increasingly competitive environment. Researching these supply networks from the perspective of workers has helped us to identify the hierarchical structure of the chains, how the structure of the chains is changing and what this means for workers.

This chapter reports on the research findings we collated from nine different garment-producing locations. As outlined in the previous chapters, this research was undertaken by local women workers' support organisations to meet their own needs, while also contributing to the international-level aims of the project. Such work can never fully represent the situation experienced by garment workers in every location, but the findings illustrate many of the challenges facing workers and their communities. They also expose the operation of those less investigated reaches of corporate supply chains. While the findings reinforce many well-known aspects and characteristics of the garment industry, they also provide insights that have not been so readily unveiled by more top-down research and less engaged investigation.

There are two sections to the chapter. The first section looks at the operation of supply chains in the garment industry. The second then goes on to explore the impact that the present structure of the global garment industry has on workers' lives. Seven key issues are identified: the employment of vulnerable workers; insecurity of work; underpayment of wages and social welfare; overwork and underwork; health and safety issues; harassment; and the challenges facing trade union organisation.

Weaving the Web: The Operation of Supply Chains in the Global Garment Industry

The research findings reflected the many changes that have taken place in the garment industry over the past thirty years. Intense international and local competition in the garment industry has meant that manufacturers in industrialised countries are outsourcing more production to lower-wage economies. Such competition has also resulted in increased pressure on several fronts, but particularly over time and price. As a consequence, in each location researchers found that there had been a dramatic increase in subcontracting. Complex supply chains had evolved everywhere as manufacturers subcontracted out production to those who could work for lower prices with quicker turnaround times.

Subcontracting splinters the industry by playing one firm, and one group of workers, off against another, all in the context of an uneven landscape of labour regulations and practice. In reality, the only way in which smaller subcontracted units are able to produce more cheaply and faster than the larger factories is on the basis of low overheads and the overexploitation of marginalised workers. Workers with few options can be employed on minimal wage rates, in poor conditions and with little or no security, and it is from this marginalisation that profits are made. Poor working conditions and job insecurity are characteristic of the industry. However, lower down the supply chain these pressures become even more intense. Those who are forced to try and survive in poor conditions at the lower end of garment chains tend to be women, many of them (im)migrants, of low caste or class.

The research revealed that workers on opposite sides of the globe had all felt the impacts of increased subcontracting. Whether they worked in India or in the UK, the effects were the same: increased temporary employment; job insecurity; irregular working hours; very low wages; non-availability of pensions, maternity leave, sick leave, bonuses or provident funds; bans of and barriers to unionism and collective bargaining; unhealthy conditions at the workplace; and sexual harassment by management. While the research revealed many overlaps and similarities, however, there were also some interesting differences. These related to variations in the regulatory context in which employment took place, the particular geo-history of industrial development in each country and the degree to which collective organisation had taken root. Such marked geographical variations were overlaid by common trends in the development of the global industry, producing a social landscape that was marked by difference *and* similarity in workers' experiences.

The structure of supply chains

Figure 5.1 illustrates the different kinds of manufacturers identified in the vertical supply chains researched. Workers in Bombay/Mumbai, for example, were able to identify 91 different subcontractors for just one firm. These included agents, as well as manufacturers/producers from Tier 1 to Tier 5. As a whole, the research allowed us to identify characteristics common to manufacturers at each tier of the garment supply chain as outlined briefly below.

Tier 1

In what we are calling Tier 1, the level closest to brand-name merchandisers and retailers, we identified large manufacturers. Often these factories were subsidiaries of multinationals whose business included textile production and manufacturing capabilities as well as a wide range of ancillary services. These firms were mostly backed with foreign direct investment, either from international or regional investors. Due to their scale and international connections, these firms tended to dominate the garment-producing industry at a national level. A clear example of this was found in the Philippines where only five firms, out of 1500 registered garment firms, controlled 20% of the garment industry in 2002. When united in employers' associations, these manufacturers formed very strong lobbies, giving them some influence and leverage over national governments, particularly in regard to changes in labour legislation and investment policies. They were also in a good position to shape the chains below them, dictating turnaround times, prices and quality levels.

These Tier 1 manufacturers provided a wide range of services, including 'full-package' production involving design, sourcing materials and

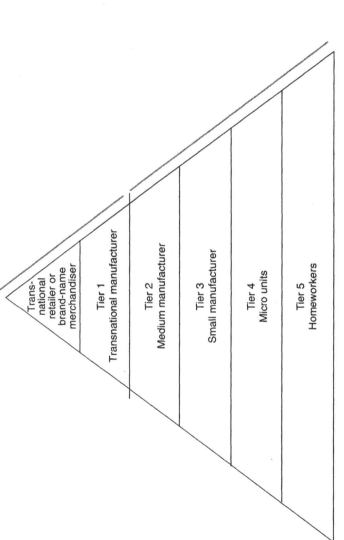

Figure 5.1 The tiers of production in garment supply chains

Source: Women Working Worldwide (2004)

Trans-
national
retailer or
brand-name
merchandiser

Tier 1
Transnational manufacturer

Tier 2
Medium manufacturer

Tier 3
Small manufacturer

Tier 4
Micro units

Tier 5
Homeworkers

distribution (for more information about the development of 'full-package' production in Mexico, see Chapter 7, this volume). As an example, the Union Research Group in India identified Go Go International as a key Tier 1 manufacturer in Bombay/Mumbai. Its capabilities extended beyond manufacturing to design, sampling, computerised pattern making, computerised embroidery, quality control, banking, warehousing and shipping. In the main, such companies produced large bulk orders for export under contract to leading brands, and produced very little for domestic consumption. Conditions in many of these factories were found to be quite good because they were often used as showcases for auditors, monitors, labour inspectors and potential buyers. Moreover, the scale and relative security of these factories made it easier for workers to organise, if regulations allowed. As an example, employment conditions at a Tier 1 factory in Thailand are outlined in Box 5.1.

Box 5.1 Employment conditions at a Tier 1 factory in Thailand

Potential workers aged from 18–28 years with Grade 4 elementary school education were recruited and tested on their ability to sew. If candidates passed the test, they would be accepted on a 4-month probation term. After probation, workers became permanent employees and received a minimum wage of 165 baht/day ($4.17). They were legally entitled to business leave, had 6 days of emergency leave and 3 months of maternity leave. Employees with longer service received higher pay in accordance with their service period. Living expenses were 270 baht ($6.82); overtime payment was 1.5 times the daily wage; the productivity reward was 3 times the daily wage during the Chinese New Year holiday.

In addition, workers had access to a number of welfare facilities, including:

Transport: Buses were provided for two routes: Bangkok and Nakorn Pathom

Food and drinking water: Drinking water was provided in the factory and canteen areas. Three pots of rice in the morning and four pots of rice at lunch were provided for the workers.

Accommodation: Dormitories for men and women were provided separately for free lodging. One room accommodated 6–7 people.

(*Continued*)

Box 5.1 (*Continued*)

Recreation areas: Annual sport activities were organized and competitions were staged during lunchtime. There were six teams and each worker was given a shirt according to their colour group. Sports included football, chairball, takraw and volleyball.

Uniform: Permanent workers received 2 shirts each year with cloth for trousers. New workers wore white shirts and each department had different colour shirts.

Health and Safety: the factory had four fire drills per year. Steel gloves were used for protection when using machinery and face masks were provided. There were no windows in the factory but fans were installed. Even so, the temperature was high. There were some 20 toilets, which was insufficient.

The right to form a union: A labour union had been formed and officially registered. Negotiations and labour relations were conducted with the employer. Several demands had been fulfilled on wages, welfare and free transport. The union now had over 500 members out of 700 employees. The committee's term of office was two years. They had co-ordinated with external organisations and other labour unions, the Friends of Women Foundation, Om Noi-Om Yai labour groups, and co-operatives. The union occasionally organised camp activities for its members.

Source: Friends of Women, Thailand.

The research undertaken in China identified a Tier 1 manufacturer that produced for Next, a large brand-name garment retailer that has stores, a home-shopping catalogue and on-line shopping in the UK. In this case, the company had contracted the order to a company in Hong Kong called the Crystal group, itself a multinational clothing manufacturer. To meet the order, Crystal group had subcontracted the sample making to a subsidiary of its own company, based in Hong Kong, and had sourced the buttons, labels and accessories through its Hong Kong office. These items were then sent to the Tier 1 factory in Guangdong Province, China, along with the fabric and thread.

In addition, however, Crystal group also used the Chinese-based subsidiaries of a number of other Hong Kong-owned garment manufacturers, lengthening the chain to include Tier 2 factories, in the production of Next's original order. Each of these factories employed about

400 to 1000 workers completing 'Cut, Make and Trim' (CMT) and packaging work. Crystal group took responsibility for the quality of the goods and ensured that the finished goods were sent to their Tier 1 factory in Guangdong Province. The researchers found that in a number of cases, these Tier 2 factories actually subcontracted some of their other work when they needed to complete the orders for Crystal group. As an example, a workshop in Tier 3 was used, employing 1000 workers to complete other work that the Tier 2 factory was already contracted to do. On occasion, smaller local workshops employing about 10 people would be used for sewing and assembly work when the Tier 3 factories could not complete orders on time.

Once the clothes had been produced in China through Crystal group, Next used its own distribution subsidiary, Next Distribution Ltd, to ensure delivery to its storage centre in the UK. From here, the clothes were distributed to shops right across the UK (see Box 5.2 for further information about Next's design and sourcing departments).

Tier 2

Amongst the second tier of garment producers, the research also identified large manufacturers, but they did not have the international scope and connections of the companies in Tier 1. In some cases, these factories were subsidiaries of Tier 1 companies, and they varied in size depending on the country. For example, in Guangdong Province in China, Tier 2 factories were found to be employing around 400–1000 workers, while in the Sri Lankan example they employed 40–80 workers at most. While the largest of these factories were often funded through foreign direct investment, most of the smaller factories were locally funded and locally owned. As might be expected, these factories worked on subcontracted orders from Tier 1 companies and also received orders directly from customers, including international brand merchandisers and retailers, as well as producing for the domestic market. Where they were part of Tier 1's chain, the factories in Tier 2 did not have much power within the chain. However, when they received direct orders from a customer, they had greater power, as they were at the top of a much smaller chain. In the main, these factories did not offer a very wide range of services and tended to focus on CMT. Reflecting their position in the garment industry supply chain, workers in these factories tended to have much poorer terms and conditions of employment than those in Tier 1. They were also less likely to be unionised and have any collective power at work.

Box 5.2 Behind the scenes at Next's design and sourcing departments

1. The marketing department presents the previous seasons' sales figures to the designers and buyers. The designers present ideas to the buyers. The buyers then determine the 'range direction' (the styles for the season) and a timetable.
2. The buyers then pick a supplier—in this case, the Crystal Group in Hong Kong.
3. The designers put together a design pack for the supplier, and the supplier produces a style sample for Next's approval.
4. If the designers approve the style and product, the buyers negotiate a price for production.
5. Merchandisers reserve fabric and agree production schedules with the manufacturers.
6. At a final selection meeting the complete range of styles is shown to directors and managers for approval before production starts.
7. Merchandisers and technologists supervise production.
8. Branch merchandisers create stock allocation targets for each style and store, based on previous sales information and predicted sales for the season.
9. A quality check takes place at the manufacturers.
10. The stock is distributed.
11. Actual sales are monitored, and adjustments are made to orders to reflect high and low sales for each style.

Source: Hong Kong Christian Industrial Committee, Hong Kong

Tiers 3 and 4

In the lower tiers of the garment supply chain, the research identified a mixture of small factories, workshops and groups of workers operating out of someone's house. In what we are referring to Tiers 3 and 4 of the supply chain, the size of these units varied depending on the location: in Sri Lanka they could have 20 employees, in the Philippines we found workshops with 10 people, while in China the scale of activity was much larger with several hundred workers in Tier 3 units. Tier 3 and Tier 4 units shared many characteristics. For the purposes of our research, Tier 4 units were defined as units operating out of the owner's house. All such operations tended to be funded by local capital, sometimes by local entrepreneurs moving up from Tier 4. None were found to have very much power in the garment industry, and their circumstances were almost completely determined by the flow of orders coming from higher up the industrial chain. Where they were supplying the international market, these companies did so as subcontractors; otherwise, they worked on CMT for the domestic market, including local retailers and/or wholesalers. Workers' rights and health and safety were under great pressure at this level, particularly for those working in the smaller workshops, as outlined in Box 5.3.

Box 5.3 Working in a neighbour's house

This case relates to a group of workers in Sri Lanka. The workers said that they were self-employed, home-based workers, producing parts of garments for Next. They were a group of seven workers who worked together in one person's home in a village about 30 km north of Colombo. Those involved in the industry were family members and neighbours.

The worker interviewed was the daughter of the woman who was considered the 'boss.' Field researchers said that this small group of workers considered themselves to be self-employed as the work was not regular—about two weeks per month—and they did not get enough work to make payments for social welfare payments or other benefits.

The workers received ready-made T-shirts with the Next logo embroidered on the front from a large garment manufacturer in the area. They were then required to hand-stitch and attach approximately 20 beads over the embroidered letters. For this work, they received 10 cents of one rupee per T-shirt, the equivalent of $0.001 (US$).

If the large garment manufacturer was not satisfied with their work, the workers were given a special pass to go into the factory and redo the beadwork. They were not paid to redo the work. The worker interviewed said that when the work was not corrected, deductions were made from their payment.

If the work was regular, workers could earn between R2000–2500 ($20–26) per month. The T-shirt in question was found to retail at approximately £15 in the UK (approximately 2,350 Sri Lankan rupees).

Source: Transnational Information Exchange-Asia, Sri Lanka

Tier 5

Homeworkers were found at the end of supply chains in every location. These workers have been included as a separate Tier, at the base of the iceberg model, to draw attention to the distinct nature of their work and their situation. By homeworkers we are referring to individual workers working from their homes, as distinct from home-based units where a few women were found working together from someone's house. Homeworking was found to vary in the supply chains included in the research. First, such work could either be supplementary, done in the evenings and at weekends, or it could form the workers' core and only income. Second, homework could be seasonal. And, third, the products were either passed to an agent, a contractor or they were sold directly by the homeworker at local markets. In this context, the nature of the homeworkers' work was also found to vary quite considerably: whereas some produced their own garments for sale, others completed particular tasks to order. Frequently homeworkers provided stop-gap production when manufacturers lacked time to finish an order, for example sewing and trimming pre-cut garments.

Homeworkers often produced goods for export as well as for the domestic market. However, they only had a relationship with the next tier up in the chain. In Sri Lanka, most of the homeworkers got work directly from a firm, while a quarter of those interviewed said they got it from a subcontractor. Several reported that they generally worked for one agent, but in quiet periods, they took work from additional sources. Many of these homeworkers knew their agents or the management in the factory, either because they were related to them or had worked in the factory previously, or because they lived in the same community. Homeworkers usually had to meet the costs of their own equipment and overheads. They had almost no power in the chain and, in most cases, they lacked the organisational structures necessary to lobby manufacturers and governments effectively.

Working conditions for homeworkers were found to be worse than in factories, with health and safety hazards affecting workers' families as well as themselves. For example, researchers found homeworkers in Sri Lanka stuffing down-filled jackets with goose feathers. This work could induce allergies and breathing difficulties, and the feathers constituted a fire hazard. The work was considered too dangerous to be done in the factories and was subcontracted out to homeworkers so that the factory continued to comply with health and safety requirements. Two examples of homeworking, from China and India, are presented in Boxes 5.4 and 5.5 (and examples from the UK are included in Chapter 6, this volume).

Insourcing and outsourcing

The research found that manufacturers adopted two different strategies to increase their flexibility and thus their ability to meet the orders that came down the chain: insourcing and outsourcing. Both methods were aimed at increasing output and are summarised in Table 5.1. Insourcing was found to involve bringing extra workers into the factory during times of peak demand or increasing the workers' overtime. The payment for this was often below overtime rates and frequently late, and sometimes the workers were not paid at all. Workers had to do overtime, either because they were asked to or because they were locked into the building and could not leave until managers allowed them out. In addition, manufacturers might use additional contract and piece-rate workers during peak seasons and these workers were either hired directly by the management or employed through a recruitment agency. If they were hired through a recruitment agency, the workers were not paid by the company, and it had no legal responsibility for them. In the main, such workers were paid by piece rate and were not entitled to any social welfare benefits. As an example, the research in Thailand found

Table 5.1 Insourcing and outsourcing to increase production

Insourcing	Outsourcing Distribution through:	To:
Getting workers to do overtime	Management	Independent companies Related companies Homeworkers Agents
Hiring additional workers during peak times	Agent	Other agents Smaller companies Small home-based units Homeworkers
	Line leaders & Supervisors	Other agents Small home-based units Individual homeworkers

Box 5.4 An example of homeworking in Guangdong Province, China

Auw was a married woman aged 33. She had been working as a garment homeworker for four years. Her 'employer' was a man called Mr X living in the neighbourhood. He produced and sold garments in the domestic market. Mr X bought fabrics from the wholesale fabrics market and cut them at home. He did both sewing at home and distributed the cut materials to homeworkers. Mr X collected the finished products from the homeworkers and sold them to small clothes shops in the towns nearby or in the garment wholesale market in big cities. Garments for the local market were low-priced, low-quality products that were sold at around RMB10-15 (US$0.83-1.00).

Auw had thought that working at home would be easier and freer. It was not. The income was not as high as she expected. She had to pay for the sewing machine, electricity and threads. If she bought the wrong colour thread, she had to pay for it. Because of the pressure and the limited amount of income she could make, Auw was looking for a job in the factory. It would save her the investment on tools and materials. There was also less pressure and the working hours were more regulated. The only disadvantage was that she could not take care of her family, and particularly the children, while working in the factory.

She reported that the contractor knew how to 'maintain' the stability of the home workers by the distribution of orders. He placed fewer orders with homeworkers who were 'difficult,' not quick enough or less skilled. Because the income of the homeworkers was controlled by the size of the order placed by the contractor, he could exert a lot of control over them. In the peak season, for example, the homeworkers had to finish a lot of orders within a short period of time. They knew that if they could not finish the orders on time, they would be penalised by getting fewer orders from the contractor next time. The rest of the family were thus recruited to help out. Not least, because the home workers had too little work in the quiet season and didn't want to upset the contractor and lose potential orders in future.

Source: Hong Kong Christian Industrial Committee, Hong Kong

Box 5.5 An example of homeworking in India

Vasu, a homeworker interviewed in India, worked mostly on thread trimming. She reported many grievances. Piece rates were set at a level that made it impossible to earn anything like the minimum wage, so other members of the family, especially girls, had to work too. Apart from this obvious problem, Vasu also complained about hidden deductions. For example, she had to buy the thread-cutter for trimming, and thread if she was buttonholing, which cut into the piece rate. Moreover, she had to carry the heavy bundle of clothing from the factory to her home in a gunny sack on her head, and then back to the factory and up several flights of stairs when the work was finished, and this work, of course, was not paid. Added to that, even the meagre wages were not always paid on time. If she arrived late at the factory there might not be any work, and she would have to go home empty-handed. The same problem arose when demand was slack.

Despite this lack of security, however, when there was an urgent order to be completed, the employer expected homeworkers to put everything else aside in order to complete it as fast as they could. As Vasu explained: 'The employer doesn't consider our problems...if we can't complete an urgent order, he yells at us or threatens not to give us orders in future and we have to tolerate this behaviour.' At such times, Vasu found it impossible to combine her work with her domestic duties. Not only did homeworkers not get any paid leave or holidays, they were also often forced to work extra on festival days, when they would have preferred to relax with their families.

And on top of it all, the supervisors and male workers at the factory often made 'dirty remarks' to the women, making them feel they were not treated with respect. Along with her companions, Vasu felt that there should not be discrimination between factory workers and homeworkers, as she put it: 'We are workers, and there should be equal respect for our labour.'

Source: Union Research Group, India

that these workers, who were made to do night work, were paid less than, and lacked the same rights as, the permanent staff.

When the management outsourced production, work was sent out of the factory to be completed by other workers. This was also done in several different ways. Managers might send the work to another company or use homeworkers. But in some companies, the line leader[1] or supervisor also acted as an agent/subcontractor and gave the work to home based workshops and/or homeworkers in the local community (for further elaboration, see Box 5.6). Both insourcing and outsourcing increased divisions and tensions between workers, within and beyond the factory. Insourcing led to sharp differences between permanent and temporary staff, and outsourcing created divisions between different production units, agents and individual homeworkers. Although they might be working on goods for the same garment industry supply chain, these workers had very different experiences and conditions of work.

Box 5.6   Employee or employer? Subcontracting to family and friends

During peak seasons, line leaders in large manufacturing units in the Philippines were found to subcontract out work to homeworkers or to another subcontractor. Researchers discovered that these contacts only paid homeworkers 20-30% of the price they received for the tasks. The home workers earned about P88 (US$1.64) per day. In contrast, one line leader who had 80 homeworkers during the peak season was able to 'earn' US$316.80 per day.

Line leaders acting as subcontractors showed that the dividing line between being an employee and employer can sometimes be hazy, creating tensions at work and in the community. Management allocated additional quotas to line leaders they liked. Researchers found that some line leaders had enough work to subcontract out to four or five homeworkers while others got enough to subcontract out to as many as eighty homeworkers. The allocation of this work within the factory, and then within the community created hostilities and tensions between people. Factory workers and homeworkers saw the line leaders as a source of additional income, but they also recognised the exploitative nature of these relationships.

Source: Women Workers' Project, Philippines

Power relationships within the chain

As already outlined, the way in which the supply chains of the garment industry operate increases tension between those involved at different stages of the chain. There are clear power differentials involved. However, the way in which power moves is extremely complex, not least because most company supply chains overlap with others. Looking at any one manufacturing company, researchers found that they were likely to be embroiled in a number of different supply chains, supplying a number of different brand names and retailers, each of which had different power dynamics. Mapping out the power relationships was thus rather tricky in practice.

Nevertheless, the participants in Bulgaria were able to develop a form of categorisation to explain what they saw as the differences in the power relationships between the firms they researched. They chose to use the term 'network' rather than chain and, building on the work of McCormick and Schmitz (2002), explored four different sets of relationship. 'Balanced' networks were seen as characteristic of mass merchandisers and involved partnership and negotiation between firms of more or less equal power. In contrast, 'hierarchy' networks were controlled by well-known brands and private labels with high levels of vertical integration and little power at the local level. 'Directed' networks, associated with designer labels, were managed even more closely, with the lead company dictating exactly what was to be produced. 'Market' networks on the other hand, involved an arms-length relationship between firms, little communication or co-operation and no long-term commitments. These were seen as characteristic supply chains for low-priced goods and discount stores. The Bulgarian researchers used the links between the multinational branded retailer, Induico, and its operations in Bulgaria as an example of a 'balanced' chain where benefits were relatively mutual; relationships with Fanco SA as a case of a hierarchial network where local manufacturers had very little power; and those with Benetton as an instance of a directed network. Finally, they explored market-based networks that tended to involve smaller and medium-sized companies producing for a range of customers and markets. In each case, they not only mapped out the networks but also looked at workers' rights within the workplaces supplying the different companies involved.

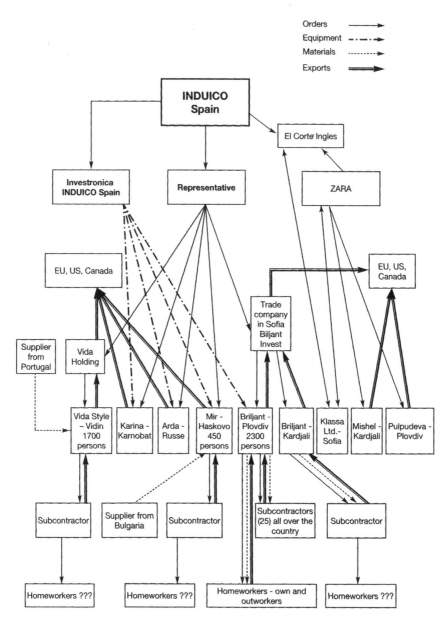

Figure 5.2 The Induico supply chain

A 'balanced' network

Zara is owned by Inditex, one of the biggest Spanish merchandisers/manufacturers in clothing. It supplies both Inditex and (see Fig 5.2) Induico via Induico's subsidiary El Carte Ingles. The researchers found that the Bulgarian companies producing for Induico, Inditex and Zara reported having a fairly balanced and co-operative relationship. Zara was working directly with these Bulgarian manufacturers to provide the designs, but they allowed the manufacturers to choose the fabrics. All the manufacturers in the chain were assisted to upgrade through equipment leases and technical assistance offered by Induico's subsidiary Investronica Induico.

At the worker level, however, the situation was found to be different. In most of the companies, the workers were on short-term contracts and their workplace relationships felt quite insecure. On the positive side, workers were given breaks, and unions existed. But, on the other hand, workers often had to do overtime and their unions had trouble renegotiating collective bargaining agreements and improving conditions. The supply chain is represented in Figure 5.2.

A 'hierarchical' network

Fanco S.A. is amongst the 30 largest knitwear production companies in the world, producing an estimated eight million garments a year. At the time of the research, it had four manufacturing units in Bulgaria and these were run by Greek nationals. Greek-based factories supplied pre-cut materials and accessories which were sewn in Bulgaria and then returned to Greece, with 'Made in Greece' labels.

In these Bulgarian factories, the workers had fixed-term contracts. In some cases they hired registered unemployed workers, so that the employers could save on salaries as the government paid the minimum salary to these workers while the employer only paid their social security contributions. There was overtime on a regular basis even on the workers' day off. In one enterprise, work was done in two shifts and in another factory, it was done in three shifts. The workers' pay was €100–129 a month (US$126–162), which was below the national average. Pay was calculated on a complex 'fixed minimum + time rate + piece-rate' system, which was unclear to the workers. In another factory, workers complained of a repressive atmosphere, restricted access within the factory and a constant air of suspicion. This chain is illustrated in Figure 5.3.

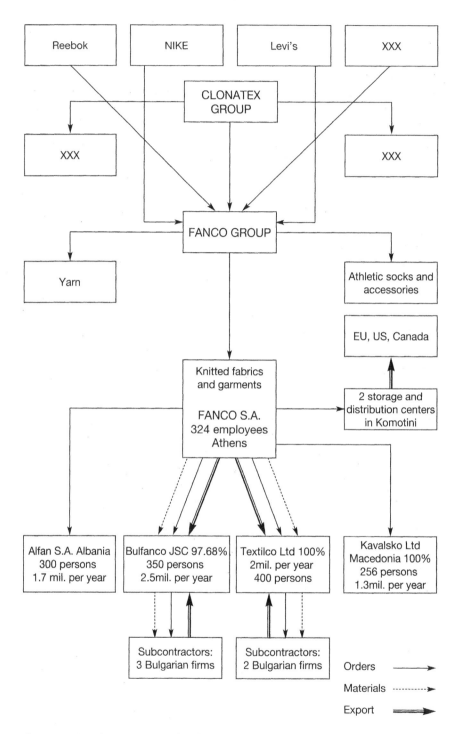

Figure 5.3 The Fanco supply chain

A 'directed' network

Production for the Italian company Benetton is organised around a 'pole model.' In this model all the manufacturers have their work allocated from a central co-ordinating office that is either totally or partially owned by Benetton. In order to take advantage of labour-cost differences, these networks have been extended into Central and Eastern Europe. Our research found that Benetton Hungary co-ordinated the production activities of contractors in Hungary, Moldavia, Romania and Bulgaria. Bulgaria was Benetton's third-largest supplier after Italy and France. The researchers found that Benetton placed their orders with an agent who then sent the order to four different manufacturers in Bulgaria, each of which also produced for other brand names. The material and accessories were supplied by third parties. Some of the factories subcontracted the work out to smaller units, as and when they needed to do so.

These four manufacturers were all Bulgarian-owned and had been privatised after the fall of communism. The workers had permanent contracts, and conditions were described by the workers as 'good.' Overtime was done on Saturdays and in some of the enterprises the workers were paid the correct amount for overtime. Homeworkers were not used by these companies, but it was not clear whether or not the smaller subcontractors further subcontracted the work to homeworkers when required. Again, this chain is illustrated in Figure 5.4.

A 'market-based' network

The market-based network accounted for the largest number of factories making up orders for export, supplying nearly half of Bulgaria's total garment exports. They were predominantly small and medium-sized enterprises producing a variety of products for small orders, with tight deadlines and under great pressure. These manufacturers were spread throughout Bulgaria, although certain ownership patterns did emerge. Inland, Bulgarian or foreign-owned manufacturing units dominated; in areas of high Turkish citizenship, Turkish owners dominated—or Turkish-Bulgarian partnerships; and near the Greek border, Greek or Greek-Bulgarian ownership dominated. In most cases, the clothing was stitched and returned to the country from where the order came. However, in the case of orders from Turkey, these orders were sent directly to Western Europe, the US or Canada so that the Turkish producers could

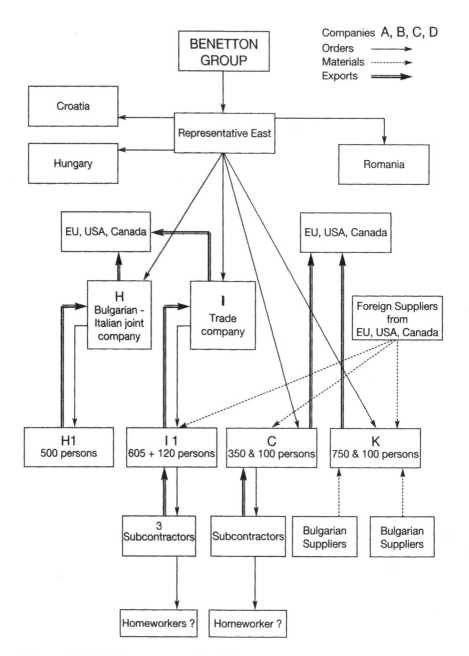

Figure 5.4 The Benetton supply chain

circumvent quota restrictions. For an example of these networks, see Figure 5.5.

The market-based networks used the highest proportion of informal labour, and production sites were difficult to trace because the sites were

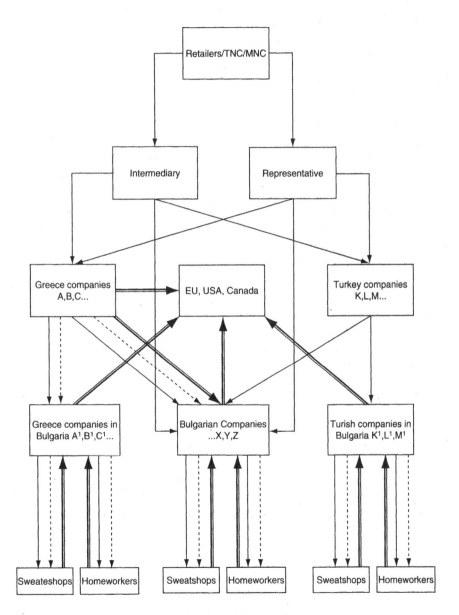

Figure 5.5 Diagram of a 'market-based' network

often not at the same location as the place of registration. Violations of labour legislation were more frequent in these networks, and there were no trade unions or other support organisations for workers

Caught in the Web: Supply Chains and Workers' Lives

This section looks at the various challenges that the structure and operation of these supply chains create for workers. The highly competitive and volatile nature of the industry creates significant pressure on workers. This pressure is shifted onto those who are least able to resist it, particularly those workers at the bottom end of the chains. Their jobs become more insecure, their wages decrease and their hours increase. The research highlighted the way in which long-standing problems have intensified and become more pervasive over time. Seven key issues emerged from the research: the exploitation of inequalities; job insecurity; underpayment of wages and social welfare entitlements; under- and overwork; health and safety issues; harassment; trade unions and organising. Each is considered below.

Exploiting inequalities

Employment in the garment industry is characterised by insecure, unstable work and a marginalised workforce. It is no coincidence that the garment industry is dominated by women. In an industry that is highly conscious of labour costs, women are generally cheaper to hire, and it is widely believed that they are also easier to control. However, the research also showed that gender inequalities were not the only inequalities that were exploited in the quest for cheaper labour. Age, ethnicity and (im)migration status also emerged as key factors.

Even though the garment industry employs more women than almost any other industry, the research found that there was a notable lack of women further up the hierarchy. While men were concentrated in ownership, management and administration, women dominated in production. In Bangladesh, for example, workers reported that it was very difficult to get any training to move into supervisory roles. The research from Bulgaria also demonstrated this bias very clearly, as outlined in Box 5.7.

As a result of the gendered nature of the garment industry, any changes in it have a significantly gendered impact, which often further

Box 5.7   The gender composition of the garment industry in Bulgaria

- Owners—10% women
- Management—20% women
- Administrative personnel—50% women
- Workers in production—90% women
- Seamstresses, including homeworkers—100% women

Source: the Bulgaria-European Partnership Association, Bulgaria

undermines women's position. The research from the UK provided a clear example of this process. The shift in the UK industry towards sourcing overseas has resulted in an increase in female unemployment. While men were employed in management, cutting and distribution, women dominated in sewing, and this work has been outsourced overseas. In contrast, management and distribution roles—where the workers were predominantly male—remained in the UK. As a result, there has been a 50% reduction in female employment in the industry over the past few years (for further information, see Chapter 7).

However, the research also highlighted the way in which gender was not a simple divide. In Pakistan, for example, stitching was often considered male work and women were hired to help the men with trimming and other tasks. In the large factories, men were employed to sew, and women worked in packing, quality control and clipping. It was only in the smaller units where women were able to sew. In contrast, women were able to work as seamstresses in the largest factories and EPZs in India, but here, men were more likely to be employed stitching in the smaller units. In each case, such divisions were further overlaid with differences of ethnicity, caste, religion and age, which also made a difference to the work that was done.

All eight reports from Asia found that employers showed a preference, especially in formal employment, for younger workers, and the preferred age group comprised those of 18–26 years. In contrast, in Bulgaria it was predominantly women aged between 40 and 50 who worked in the garment industry. This was because they were seen as more skilled and productive and because younger workers were emigrating. Homeworking, which was much more insecure, also tended to be done by older women, frequently because they had additional unpaid, domestic responsibilities.

Table 5.2 The differences between local and migrant workers in Guangdong Province, China

	Internal Migrant Workers	Guangdong-Origin Workers
Age/marital status	Unmarried young women, aged from 16 to 30.	Middle-aged married women that had children and family to take care of.
Residence	Stay in the factory dormitory	Stay at home outside the dormitory
Attitude to overtime	Willing to work long periods of overtime	Did not want too much overtime due to family commitments
Residential status	Temporary residence in Guangdong. Must get a job to stay in Guangdong. Not entitled to social security provision in Guangdong.	Residence in Guangdong province. Entitled to social welfare provisions.
Workplace	Mostly worked in foreign-owned enterprises	More commonly found working in locally owned private enterprises or as home workers
Attitudes to conditions of work	Highly tolerant of the labour abuses in the factory.	Less tolerant of abuses in the factory.

Source: Hong Kong Christian Committee, Hong Kong

Another social dimension that was highlighted in the research reports was the employment of migrants and (illegal) immigrants in the garment industry. The research from the UK revealed that illegal male immigrants were finding work in the UK garment industry. Their illegal immigration status ensured that they would not complain to the relevant authorities if they were underpaid or if other rights were violated. The research from China also revealed significant differences between internal migrants—who lack the right to social welfare once they leave their province—and local workers. As a result, the migrant workers were more willing to accept terms and conditions that more secure local workers would refuse (see Table 5.2).

Job insecurity

The research clearly showed the increasingly informal nature of the industry. In all nine countries where the research was carried out, the industry was dominated by small-scale factories and workshops, frequently employing workers on short-term contracts or without any contracts at all. In Thailand, for example, 70% of the labour force were employed in small and medium-sized units where conditions were steadily declining. In Pakistan, 70% of units had 4–10 machines and two irons. In the UK, the informal workforce was found to have increased, while the formal workforce had decreased, and the informal workforce was estimated to be only 20% smaller than the formal workforce.

The research also highlighted the trend towards increasing informalisation within the formal sector. In the Philippines it was found that in one firm employing 700 workers, only 160 had permanent contracts. In every case, the increase in temporary employment in the formal sector acted as a silent warning to workers with permanent contracts. They understood that their situation was precarious and that there were other workers available to take their place if they stepped out of line. Moreover, the range of different employment practices deployed even within any one factory increased workers' confusion about their status at work. A clear indication of this was found in the research in Sri Lanka. There, some of the workers interviewed described themselves as 'self-employed' even though their hours, pay and work were clearly organised by someone else. Designating workers as self-employed means that employers can avoid paying the workers' social welfare contributions.

Changes in government legislation have also changed the structure of work. Under communist rule in Bulgaria, all employment was formal employment and this remained a strong legacy in the post-communist state. The researchers found that employers had begun lobbying the government to ensure that employment in the garment industry was classified as 'seasonal' or 'temporary,' and court decisions in Bulgaria supported this interpretation of the legislation. As a result, these changes undermined the rights of workers. In many places it was reported that labour laws did not apply to small companies or those without contracts, so that workers were unable to protect themselves. In India, Pakistan and Bangladesh, over 90% of those interviewed had no contract letter. This meant that they could not prove that they were workers. Consequently, they were ineligible for certain legal entitlements. If they were fired, they could not prove that they were ever employed, so employers could escape any responsibility for their staff.

As work has become more insecure, many women have been retrenched and have had to work from home. Homework was frequently done by older women, many of whom had children and needed to balance income generating with childcare. In the UK, Asian women form a large proportion of homeworkers owing to language barriers, childcare, lack of social networks and restrictions on working outside the house (for further information, see Chapter 7, this volume). In Guangdong province, China, monthly incomes for homeworkers fluctuated from RMB800 (US$66) to RMB3000 (US$250), depending on the season. Moreover, these homeworkers had to make a down payment to the agent who gave them work. This was generally a month's salary in advance to ensure that the homeworkers would complete the work for this agent and not sell the goods to anyone else. The money was returned to the homeworkers when they ceased to work with that agent. Homeworkers in all countries reported that they have to pay for their expenses, despite earning so little from the work that they do.

Underpayment of wages and social welfare entitlements

The most common complaint made by the workers interviewed for our research projects was that wages were low, late and incomplete. The research found that wages were often purposely made complex, so the workers could not calculate their wages in advance and did not know if they had been underpaid. Wages could be based on piece rate, time rate, minimum wage or a combination of all three methods. Figure 5.6 shows the ways in which wages and wage-security differed along garment industry supply chains. Compiled from the research in Guangdong province, China, these conditions were rather better than they would have been in other provinces in China. In Guangdong province, the minimum wage was RMB450 (US$37.50) per month, and this is higher than that found in other areas of the country.

This example illustrates several features that were common to the research findings in all nine countries involved. First, wages tended to be higher the closer the factory was to the top of the chain. Second, bigger factories usually had a minimum wage for their permanent workers during low seasons, but most workers were made redundant, frequently without compensation. Third, piece-rate payment was almost universal, particularly outside the big (Tier 1 and 2) factories. Workers who worked on the piece-rate system experienced dramatic changes in income between high and low seasons. Finally, late payment was not uncommon and in some cases, payments were reported being as much as three months late.

The problem of sub-minimum wages appeared in all the reports. In Bulgaria, the women interviewed were often the main breadwinners.

First Tier - Supplier Factory

- RMB 700–1000 (US$58–$83) a month in peak season
- Piece rate
- Minimum wage paid in low season

Second Tier - Subcontractor Factory

- RMB 500–1000 (US$41–$83) a month for skilled workers in peak season at piece rate
- RMB 300 (US$35) a month in low season
- Piece rate
- Wages paid 2–3 months late
- No minimum wage or subsidy in low season

Third Tier - Subcontractor Workshop

- RMB 500–800 (US$41–$66) a month for skilled workers in peak season at piece rate RMB 200–300 (US$16–$35) in low season
- Piece rate
- No minimum wage or subsidy in low season

Fourth Tier - Subcontrator Units / Home workers

- RMB 800–900 (US$66–$75) a month in peak season at piece rate
- RMB 200–300 (US$16–$35) in low season
- Piece rate
- No minimum wage or subsidy in low season

Source: Hong Kong Christian Industrial Committee, Hong Kong

Figure 5.6   Comparative wages in a typical supply chain in Guangdong Province, China

During the interviews many women began to cry when talking about their wages. One woman got a salary of €50 (US$62) of which she spent €21 (US$26) on transport and the remaining €29 (US$36) had to cover her family's costs for a month. In Pakistan, although 95% of the women interviewed were earning supplementary wages, they began working because their families were in financial crisis. The minimum wage in Pakistan for unskilled workers was R2500 (US$43) per month (or US$1.43 per day), but they were paid less than this by employers. In Sri Lanka, workers told researchers that 'the earnings were not sufficient for food,' while another woman declared that: 'after I pay for my food and boarding, I am not left with anything to be sent to my mother.'

Equally, in the UK, women workers in the garment industry earned about 33% less than men in the same sector. In the informal sector they earned as little as £1 per hour (US$1.50), just 25% of the national minimum wage of £4.10 (US$6.75) per hour at the time of the research. In Bangladesh, workers were legally entitled to allowances for accommodation, transport, meals, medical bills, pension, provident fund and insurance, but they were often unaware of these entitlements and few received them. In the Philippines, transport and living allowances were not paid, and many workers were unaware that they were entitled to any such benefits. In Guangdong province, China, payment of social welfare contributions only occurred in Tier 1 and Tier 2 companies. These payments were only made for high-status workers, normally managers and senior administrators, but not for all manufacturing staff. As management was almost exclusively male, while manufacturing workers were predominantly female, there was a significant gender bias in social welfare payments.

Coupled with underpayment of wages and retention of legitimate allowances, workers had money deducted from their earnings. They typically faced a long list of offences for which money was deducted from their wages. These ranged from being sick or late to not achieving daily or weekly quotas. In some cases, legitimate deductions for various social welfare payments were made, but these deductions were not always remitted to the government. Quality control was a particularly contentious area in relation to deductions. In some cases, workers were obliged to repair errors without payment. In other cases, workers had money deducted for each mistake and also had to repair the mistake without pay. In Guangdong province, China, the researchers found a factory with very strict quality control, where workers were fined RMB10-50 (US$0.80 - $4) for each defect. This fine was not shown on the workers' pay records, so it was not seen by external auditors coming to look at the site. Combined with complicated wage structures, these deductions made it almost impossible for workers to calculate their wages in advance or to check that the amount they were paid was correct. Our researchers found that many workers were forced to sign blank dockets when they collected their wages. In this way the employers could write a higher wage on the docket for monitors, should they come to inspect.

Hours of work

The research confirmed the polarisation of working hours, with many workers experiencing periods of over- and underwork. During peak seasons, workers were expected to work all the hours necessary to finish an order, and on average the peak season was about three months long. The researchers found that workers had to be at the workplace 7 days a week for this time. During low seasons, however, redundancies and the forced use of holiday leave were commonplace.

In addition, working days varied in length. While overtime in the large factories was officially limited to two to four hours per day, the hours worked in other factories and smaller units were often much longer, and twelve to sixteen-hour days were not uncommon. Eight-hour shifts with two hours overtime were mentioned in the reports from Sri Lanka and Bangladesh. In Guangdong Province, China, ten-hour shifts with two hours overtime were reported and twelve to sixteen-hour shifts were noted in Pakistan. Most workers were denied their day off during the peak season (see Box 5.8 for further illustration of the problems caused by irregular hours of work in Pakistan).

The extent of overtime working seemed to depend on the factory's location in the chain, with better conditions being more likely towards the top of the chain. These factories were more likely to monitored and audited; trade unions—where they existed—had sometimes negotiated less overtime here; and some large factories used shift work to allow workers to leave their positions so that the next shift could begin. As might be expected, the willingness to commit to overtime was closely related to inadequate wages. In Bulgaria, a living wage for one person was estimated in 2002 as €162 (US$138) per month, but the average salary of a seamstress was about €115–€130 (US$98–$111). To compensate for the low pay levels, these women would work 12–16 hours a day when they could. In this way they could earn up to €225 (US$192) and so provide something for their families. On average, the workers interviewed in Bulgaria did 70–150 hours overtime every month and frequently worked seven days a week when possible.

The more informal the work, the more 'flexible' the hours worked were found to be. Greater industrial flexibility means wider fluctuations in the orders sent to particular factories, and greater insecurity for the workers involved. One of the clearest demonstrations of this was the situation facing homeworkers in Sri Lanka and the Philippines, where researchers found that 16-hour days were not uncommon and some workers did as much as 60–90 hours a week. This figure did not include the additional work carried out by members of the homeworkers' family, either with direct help fulfilling the quota or with assistance in completing domestic chores.

As the flip side to this problem, however, underwork was also reported. Just as breaks and leave were inadequate and difficult to take during peak seasons, redundancy and the compulsory use of holiday leave were commonplace during the low season. Workers were obliged to use their holiday leave during the low season so that the company did not have to pay them to be at home when orders were high. And these

---

Box 5.8   The problem of irregular hours of work

Rani worked in the finishing department of a Knitwear factory located in Lahore, in the province of Punjab, Pakistan. At the time of the research, there were 15 women and 5 men in her department and 500 women in the factory in total (most aged between 14 and 30 years old).

The company was exporting T-Shirts and jeans to the US and UK. Rani had been working for three years, but she was still a temporary worker. She started her work at 7:00 a.m. and finished at 10:00 p.m. or 11:00 p.m. She had no fixed working hours, so she had no idea when she would finish work. She told us: 'When boss allows us, we go home. We work long hours without getting overtime. My male supervisor misbehaves and harasses me by passing unwelcome remarks. He tried to start affairs with the younger girls and threatens that he will stop their salary if they refuse. If I refuse to work overtime, the supervisor and manager would ask me to leave the factory. We are not allowed to talk with each other during work.'

Rani's factory had no separate toilet for women. There was no place to eat so the workers sat on the floor at lunchtime. There were no fixed break times, so the workers weren't always given their breaks. The lights were dim, and Rani suffered from severe headaches and eye problems. There was no proper air ventilation system, and because of this most workers suffered from asthma and other lung problems. Pregnant women did not get any maternity leave.

Payment was also a problem. Rani was not getting equal wages for equal work. She worked on a piece rate and got R1200 per month (US$ 24). Her employer gave her the salary and made her sign on a piece of blank paper.

Rani told us that there was no union in the factory. In the past, if any worker tried to form a union he/she was fired without reason.

Source: Working Women's Organisation, Pakistan

---

were the lucky workers, for in every location researchers found that significant numbers of workers were simply made redundant during the low season each year.

Health and safety issues

Rani's situation in Pakistan (as outlined in Box 5.8) illustrates the way in which health and safety was a recurring theme in the research. The issues highlighted were almost identical in every location. They included excessively high temperatures—or very low temperatures in the case of Britain—dust, inadequate ventilation, inadequate lighting, excessive noise, lack of fire-fighting equipment, blocked exits, bad sanitation, unhygienic canteens and lack of drinking water. The physical effects of these conditions were exacerbated by sitting bent over a sewing machine on stools and broken chairs, or using a heavy iron all day. The list of illnesses, infections and injuries cited in all the research reports were almost as long as the list of health and safety violations. They included fevers, headaches, eyesight problems, skin allergies, kidney infections, backache, stomach cramps, breathing difficulties and constant exhaustion. The report from Thailand found that in some factories, separate dormitory space was not provided for workers, so they even had to sleep in their workspace.

Toilet breaks were generally inadequate, and some companies did not provide drinking water in order to minimise the number of toilet breaks taken. Researchers in Sri Lanka found that workers were only allowed one-minute toilet breaks in some factories. In other countries, workers were only allowed two toilet breaks in a ten or twelve-hour day and even these might be timed. The result of this was dehydration and kidney infections. The report from Bangladesh found that women spent 8% of their salary on tackling their ill health but men spent only 4%. It is possible that this is because women are more likely to work in production, where health and safety standards are lowest. Despite the conditions in which women had to work, it was difficult for them to take time off due to illness, not least because they very rarely had payment for these periods, and most workers continued to work even when they were sick.

Many employers were unwilling to invest in improving health and safety standards, either because it would cost too much, or because the units were rented so they did not see the point in investing in the buildings. The Bulgarian researchers found that according to the Labour Inspectorate's estimates, about two-thirds of the buildings used in the industry were rented, and these factories generally failed to meet minimum standards. Some of the problems facing workers in the Bangladeshi factories are outlined in Box 5.9.

Box 5.9   Behind the factory door: Health and safety in Bangladeshi factories

We observed, during our factory visits, that most of the garment factory buildings were overcrowded, congested and poorly ventilated. As a result, garment workers, particularly female workers, were exposed to toxic substances and dust. Raw materials used in the garment industry include various types of fabric, cotton, synthetic and wool, which contain dust and fibre particles. Clouds of thread particles hung in the air. Dye is a toxic substance emitted from coloured cloth, and it spreads in the workroom. The workers, particularly the operators and sewing helpers, who are mostly women, continuously inhale these substances. Congested and overcrowded working conditions, without proper ventilation, result in temperature hazards.

Garment workers were exposed to fire hazards too, as most garment factories did not have adequate fire prevention measures despite the fact that garment factories are very prone to fire accidents since all the raw materials used in this industry are highly flammable. Many participants said that most of the fire-fighting equipment did not work and that it was just there for show. Fire exits tended to be very narrow and very dangerous. Also, in some factories, fire exits were used for storing boxes. Participants said that most of the deaths caused by fire in the garment factories were due to accidents while running down the stairs, and not the fire. Most workers die in the stampede.

Source: Karmojibi Nari, Bangladesh

Harassment and violence towards women

Three different types of harassment were identified through the research: verbal/psychological, physical and sexual. Female workers were more likely to be harassed than male workers. This harassment was only highlighted in four of the ten research reports, but it is likely that the issue was underreported. The use of physical harassment and intimidation against trade union organisers and members was highlighted in the reports from Bangladesh and the UK. In Bangladesh, workers involved with trade unions faced redundancy, harassment and intimidation, as well as being threatened with murder.

The report from Pakistan noted that over 70% of the women workers interviewed had experienced some form of work-based sexual harassment. This included prohibitions on lipstick, make up and henna, and supervisors forcing women to cover their heads. Many were not allowed to talk to any men, yet, in all the factories visited, women had only male supervisors. In Bangladesh, half of the women surveyed had faced violence and harassment in the month prior to interviews.

The reports from Pakistan and Bangladesh also highlighted a link between insecure jobs, underpayment, overtime and harassment. The women needed to work overtime at their jobs when asked to, either because they feared losing their jobs or they needed the money. When women were allowed to go home at the end of the overtime shift, it was often very late at night and there was no secure transport available. As a result, they had to walk home or take unsafe transport. The research noted that women workers were more likely than male workers to be the victims of robbery, physical attacks, beatings, kidnapping, sexual harassment and rape. Zarina's and Delowara's stories (Box 5.10) illustrate the devastating impact these incidents have on women's lives.

Box 5.10   Zarina's and Delowara's stories

In May 2001, I along with six women workers was kidnapped while going home. The kidnappers took us to a building site where six persons raped us and threw us in the back of a wagon. In spite of not giving any compensation to my other colleagues and me, the factory management refused to accept us as their employees. My family and I approached the local police station, but the police attitude was very rude, and no criminal case has been registered against the culprits. That is why I left my old company.

(*Source*: Zarina's interview with Working Women's Organisation, Pakistan)

One day, a local tout, who was teasing me for a long time, forcefully took me on a scooter while I was returning home after finishing 5 hours of overtime work at a garment factory, which was about 2 kilometres away from my residence. I was crying loudly, but nobody came to save me. The tout put a handkerchief on my mouth, and within a few seconds I fainted. He took me to his relative's house, where he identified me as his wife. He proposed to me to marry him.

(Continued)

Box 5.10   (Continued)

But I did not agree since he was a tout and drug addict. That night he raped me several times. I was bleeding profusely. But he did not care for that. He kept me in that house three days, and every day he raped me inhumanly. After three days he left me free. I came back to my house where there was my parents and two young sisters. My parents were very angry to see me. They did not allow me to reside with them. My elder sister, who was married and a garment worker, helped me a lot. She took me to my parents' house and reasoned with my parents that I had no fault in this event. My father went to the court and filed a rape case on my behalf. Within a few days that tout was arrested. He is in jail now. My father also took me to the private doctor. He requested the doctor to keep everything secret.

I was accepted in my family, but the neighbours did not accept me. The neighbours looked down upon my family and me. They often passed demeaning words at me and at my parents. It became impossible for me and for my parents to live in that area. My father was forced to shift our residence to a new place where nobody knows about my accident. In this place we are living in peace.

Not only the neighbours, but my colleagues also did not accept me. They always laughed at me. The male colleagues were calling me a prostitute and invited me to their bed. The female colleagues were avoiding me. They did not talk to me. I could not stay in that factory. I took a job in another garment factory where I changed my name. Here I am working as a finishing helper and drawing a monthly salary of Tk1000 including overtime income (US$21).

I had got back my family and job. But I had lost my identity. Now I am a new person. I don't have any future. Nobody will marry me. My parents are trying to get me married. I will marry that person who will agree to marry me after knowing the violence happened against me. But I know nobody will agree. I know I have to lead my life unmarried. Now I don't feel interested in anything. I don't have any aspiration. Previously I used to go to cinema. But now I don't go to cinema. Always I feel that people are laughing at me. Now, I am always afraid of being raped either in the factory or on the street. I don't know how long I have to live with this psychological condition. No rehabilitation centre or government could save me from this psychological condition.

(*Source*: Delowara's interview with Karmojibi Nari, Bangladesh)

Trade unions and organising

The research confirmed the increasing pressure that trade unions are facing. Subcontracting allows businesses to break up their production locations, and the result is that many workers do not know who their co-workers are, so they cannot unite with them; nor do they know who their employers are, so they cannot unite against them. The research also confirmed that employers and governments are becoming increasingly hostile towards trade unions. This has made workers very reluctant to join trade unions where they do exist. The research also found that trade unions were under pressure internally through corruption and workers' perceptions that some unions were working to support the employers rather than the workers.

The reports highlighted the very low levels of trade union membership in the garment-producing sector. Only 10% of factories in Bulgaria were unionised. Of 185 workers interviewed in Sri Lanka, only 1.6% were involved in trade unions. No independent unions were allowed at any level in the supply chain in China. And in Bangladesh, as in many other countries, trade unions were prohibited in EPZs, and only permanent workers had the right to organise elsewhere. In the Philippines, trade unions were unofficially prohibited in EPZs. And in the UK, trade unions had failed to gain a large membership because they lacked the experience and personnel needed to organise in the new smaller companies that now dominate the industry.

Government legislation was frequently found to support employers' interests at the expense of workers' interests. In India, for example, large manufacturers were lobbying governments to change labour laws so that they could hire contract workers for permanent work. Equally, labour laws in Pakistan only applied to workplaces employing more than 50 workers and did not recognise anyone who worked less than 180 continuous days in a year. In India, as in many other countries, employers frequently subdivided their units. Instead of having a factory employing a hundred workers, they tended to manage five different units each employing twenty workers. This strategy undermined trade unions; it stopped workers uniting as they were employed in different units that might be in different parts of the city. If workers in one or two units did unite and organise for better conditions, the owner could close the units and immediately transfer the work elsewhere, thus ensuring the continuation of the production in hand. This took place at Go Go International during the course of the Indian research (see Chapter 8, this volume). The research also revealed cases of workers being black-listed for union activity (see Box 5.11).

Significantly, the research in Bangladesh found that there was little difference in unionisation between manufacturers that had direct contact with buyers, and factories that did not. Despite the fact that large brand-name merchandisers/retailers had codes of conduct that entitled workers to form unions and bargain collectively, monitors from these companies had never asked any of the workers if they had any involvement in trade unions or faced any problems in organising. The Bangladesh research also highlighted the increasing use of harassment, intimidation and violence against trade union members and organisers. In Bangladesh, factory managers often employed an armed *mastan* (muscleman) group to drive out the union. In a number of garment factories, workers seen talking to union leaders were instantly dismissed

---

Box 5.11   Blacklisted for organising in the Philippines

Melody was once a worker in K-9, a manufacturing firm in the Baguio City Export Processing Zone (EPZ). The factory, owned by a Korean national, produced bags for export. She was one of the women leaders who led the formation of a workers' union in this factory. They were able to draw overwhelming support from the majority. The issues were low wages, reduced benefits and high quota systems leading to forced overtime.

The company then filed notice of closure due to financial loss that left no option for the workers than to stage a strike. Since the company had a sister company in another EPZ in the country, it was easy for them to run away from their obligations to workers. While the labour case was still pending at the Department of Labour and Employment at the time of the research (late 2002), the workers were left unemployed and deprived of the wages and benefits due to them.

Melody spent almost two months searching for a new job. She, along with the other members of the union, was discriminated against and was not hired by any other company because she was branded as 'unionist.' Being a single parent with a four-year-old child, Melody began homeworking, even though the work was insecure and she could not earn enough to cover their basic needs.

Source: Working Women's Programme, Philippines

and workers suspected of organising were also dismissed. In one reported case, local managers had printed posters listing the names of trade union activists under the heading 'Beware of these workers.' These posters were hung at the gates of all factories in the local area, making it very difficult for these workers to get jobs (for an example from the Philippines, see Box 5.11; and see Chapter 7 for more information about the UK).

Clearly, increased informalisation and subcontracting decreases the ability of workers to organise. Workers are increasingly divided between permanent and contracted staff, between those having fixed wages and those working for piece rates, and between those who are factory-based and those who are home-based. This presents a problem in forging collective identity as many workers do not work in the same place or know where other units of the same company are located. Tensions between different workers were very clear in Thailand, for example, where some unionised workers felt threatened by those to whom sub-contracted orders were sent. Similarly, in the UK, there was resentment amongst some workers that their work was being sent abroad for completion by workers who were paid much less.

The increase in informal working and in homeworking has particular implications for union organising, as informal workers and homeworkers face very specific problems. As researchers in Pakistan pointed out, homeworkers tend to be dependent on family members and friends for a source of income. In this case, organising would not only jeopardise their work but would also destabilise their family and community relationships. Moreover, they are also very isolated and often hard to reach. The researchers found active organisation amongst some homeworkers in the UK, but this was very rare in other locations.

Conclusion

The greatest strength of this research project was that it was carried out using an action research methodology that started with the concerns, needs and knowledge of workers. While the research substantiated general theories about the garment industry, it also went further and uncovered hidden dimensions that may only be visible from the perspective of workers themselves. As a result, new dimensions were added to our understanding of the way in which subcontracting chains in the garment industry function. Among these findings were the tiered structure of subcontracting chains and the blurring of distinctions between employee and employer, seen most clearly in the case of line leaders acting as agents to homeworkers. The research found that increased worker insecurity was a common theme in all the participating countries. The findings also highlighted the disintegration of laws, codes and standards in the industry across the world. Irregular working hours, sub-minimum wages and the denial of the right to organise and negotiate were all widespread. Increased industrial flexibility had led to increased informalisation and job insecurity. While workers were afraid of losing their jobs, they were more willing to accept lower wages and poorer conditions in order to keep their jobs. When wages were low, workers had to work longer hours in an attempt to earn anything like a decent wage.

The research also highlighted the hardening of hierarchies, biases and discrimination in garment industry supply chains. Gender, age, religion and ethnicity were used against workers in order to further fragment worker solidarity. This had severe implications in a climate where trade unions were under such sustained threat. However, whilst the research illustrated the many challenges facing workers in the garment industry, the data collection also became a mechanism for facing up to those challenges. As outlined in the previous chapter, it enabled bridges to be built with the most marginalised workers and provided information for lobbying on their behalf at both national and international levels. It also formed the basis of programmes of worker education, which are vital in helping workers to organise themselves and reclaim their rights.

Note

1   The line leader is a permanent worker who supervises an assembly line of workers.

6

Coming Undone: The Implications of Garment Industry Subcontracting for UK Workers

Camille Warren

Introduction

From the time of the Industrial Revolution the garment industry has played an important role in British manufacturing. Yet nowadays most of the clothes sold in British shops are not produced in British factories and are a product of supply chains that may operate across many countries. Most large factories have closed and most retailers, agents and manufacturers have moved some or all of their production overseas to take advantage of cheaper wages. This has had profound repercussions on the structure of subcontracting chains in the UK, which now tend to involve fewer units and units of a smaller size. These changes have had major implications for UK workers, many of whom have a very uncertain future.

Intense pressure on profit margins in UK garment manufacture has been caused by macroeconomic factors, such as goods being produced more cheaply overseas, international trade agreements and a strong pound, combining with inherent weaknesses in the structure of the UK industry and an overbalance of power in the supply chain towards retailers. Manufacturers within garment supply chains have responded to this pressure by cutting their labour and production costs as far as possible, passing on the insecurity of their position to workers. The effect on workers has been a massive increase in job insecurity, a reduction in the regularity of work, increased informal labour practices, a decrease in trade union representation and often the upheaval of moving workplace and industry.

In this chapter, the implications of the changing nature of subcontracting chains for workers are explored by mapping four subcontracting chains that are illustrative of the main trends in the UK clothing industry. Although the supply chains are only sketched out in brief, they allow us to understand how the downward pressures that are exerted through supply chains can be understood as generators of instability and hardship for workers. Each supply chain highlights a different response to the increasing competitiveness of the industry, and reveals how this competitiveness has increased the diversity in the composition of the UK garment industry workforce and caused striking similarities of experience between workers in different supply chains.

This research was carried out over a period of nine months in 2002 and is no more than an outline of changing subcontracting chains in the industry. However, combining workers' own understanding of the supply chains they work in with an overview of the clothing manufacturing sector, allows us to insert the workers' perspective into the analysis of a globalised industry. Interviews for the research were carried out with company owners, workers, trade unions, homeworking organisations and academics. Information from interviews was combined with industry research and academic literature to reach a fuller understanding of the clothing manufacture sector.

Looking at subcontracting chains in the UK as part of the wider WWW project allowed for a comparison of the situation facing workers in very different economies. While the UK industry has been disappearing and has substantially reorganised itself in terms of production and market orientation, the supply chains examined in Asia were set up to produce very different products for very different markets. But, despite this, there were many commonalities in workers' experiences. As lengthening international supply chains have become the norm in clothing manufacture, traditional employment configurations that formed the basis for models of collective organising, have disappeared. Understanding the basis and nature of commonalities between workers in different locations might eventually form the ground from which to develop a new approach to collective organising and defending workers' rights.

The first section of this chapter provides an overview of the industry in Britain with an emphasis on the changes in retailing and production that have most affected the industry. In the second, there is an overview of how the structure of subcontracting in the UK affects workers. This section discusses changes in the size of workplaces, the increase in informal work practices, changes in the composition of the British workforce in relation to ethnicity and gender, wage levels and trade union membership. The research examines three industrial segments of the UK garment industry in detail: homeworking, knitting factories and companies that also produce or source from abroad. The last section looks at the factors influencing power relationships within subcontracted chains in garment production. By understanding power within subcontracted supply chains we can also understand where to apply pressure in order to try to effect change.

Industry Overview

The clothing manufacture industry is no longer strategically important to the UK economy. Clothing manufacturing accounts for 4% of all manufacturing employment but only 2% of all manufacturing output and 2% of exports (Office of National Statistics 2002). Moreover, at a time when the entire manufacturing sector in the UK accounts for only 17% of national employment, the industry clearly lacks strategic importance. As a result, there has been very little government support for and/or investment in the industry and little effort made to protect Britain's share of international trade.

The clothing industry has been declining in employment since the 1970s, and, since the late 1990s, the industry seems to have entered a new phase of employment decline and sharply falling output (Jones 2002). The speed of the employment decline is quite staggering. Between 1997 and 2002 the number of employees in the clothing manufacturing industry has almost halved from 167,000 to 83,900 (Office of National Statistics 2002). This represents a fall of 11% per annum for the industry, and there is no sign that this rate will fall off, given the ever increasing numbers of UK manufacturers deciding to move production abroad and the approaching MFA phase-out in 2005.

Reasons for the decline in clothing manufacture in the UK

The UK industry has faced a large number of external and internal factors that have caused the decline in production. The largest market for British clothing exports is the EU, which receives about 58% of all

UK clothing exports. As Britain has remained outside the Euro, the currency of its major market, exports have declined. At the same time imports are increasing, now making up at least 65% of the UK market (Office of National Statistics 2002).

Other macroeconomic reasons that have been cited for the decline include the reductions in import tariffs on goods from certain producing countries. The National Strategy for the clothing industry (see Textile and Clothing Strategy Group 2002), developed by an industry and trade union partnership gives the example of Pakistan, which, since the attacks on New York on September 11th 2001, has been allowed to export clothing duty-free to the UK. Some argue that there is an 'unfair playing field,' in which other countries are keeping high import tariffs while the EU has been phasing them out too quickly. Although most producing countries have been berating the EU and the US for being slow in phasing out the MFA (see Hale 2002), British trade unions and manufacturers argue that it should be kept in place for as long as possible.

The UK clothing industry is characterised by an extremely atomistic structure in its manufacturing sector and a uniquely highly concentrated retail sector. 74% of clothing manufacturing firms have turnovers of less than £250,000 per year. At the same time, however, the high concentration of retailers in the UK clothing market has allowed them to exploit this local production capacity while driving down the prices they pay for outsourced production (Jones 2002).

A number of internal factors have also weakened the clothing manufacturing sector: the lack of capital investment caused by the constant squeeze on profit margins, the extremely labour-intensive nature of the industry, the move to uneconomical short runs, the shortage of skilled labour and the lack of value-added production. To avoid these constraints on profit and to take advantage of cheaper wage rates abroad, many UK manufacturers have adopted strategies of 'outward processing,' where materials are shipped abroad for foreign make-up (Begg et al 2003). Manufacturers, while retaining their brand name, have in many cases given up manufacture in the UK. While some own factories abroad, many use complex international subcontracting chains of agents and middlemen to ensure that their orders to retailers are filled. In this sense, manufacturers have become more like agents, 'managing' production for retailers and taking on roles like warehousing and design that were previously fulfilled by other actors in the UK clothing industry. Where a UK workforce is kept, it is often to ensure they can do short runs, like large sizes or maternity wear or to keep retailers happy through offering quick turnaround on orders and corrections of mistakes on work coming in from abroad.

Trends in clothing manufacturing employment

Traditionally, the UK clothing industry was largely geared to producing long runs for British retailers, and so the workforce was particularly hard hit when UK retailers and manufacturers moved the bulk of their production abroad. In the National Strategy, UK retailers said they were committed to maintaining some of their production in the UK in the form of short runs and quick response, but their commitment to this seems to be lessening with time. Almost all the major retailers now source offshore, including brands such as Marks and Spencer and Doc Martins, which previously prided themselves on their 'Made in Britain' tag.

Price competition in the retail market has also driven down price margins in the UK. From 1975–1988 the real price of clothing halved, which means less profit per unit (Office of National Statistics 2002). British manufacturers frequently find they cannot produce clothes at the prices retailers want to buy them and often pass these pressures on to their workers by offering lower wage rates or insecure and temporary work. Discount retailing has decimated lower-priced retail outlets, such as wholesale and market stalls, which previously sourced largely from the smaller UK-based manufacturers. As consumers continue to shop with a keen eye to price, this puts pressure on manufacturers and workers producing the goods. This trend shows no sign of abating as the high street has seen an influx of new discount clothing retailers in the last five years such as Matalan and Primark (Hayes and Jones 2002).

The decline in the clothing industry has caused significant changes in the composition of the workforce of the UK. Traditionally, the UK industry was characterised by medium- and large-size enterprises that employed a mostly female workforce with male management (Oxborrow 1999). It tended to be a white-working-class occupation, generally regarded as unskilled and of low status. As large-scale manufacturing started to disappear from the 1970s, however, this type of factory almost ceased to exist. In the wake of this decline, a new type of clothing enterprise began to spring up during the mid-1980s that was much smaller in size and focused on either specialist areas such as technical textiles or low-value sectors such as socks, low-priced CMT (cut, make and trim) and knitwear. These enterprises have kept some employment in the UK, but the high start-up rate is accompanied by a high rate of business failure and subsequent job insecurity (East Midlands Development Agency 2001)[1]. The production units are associated with a rise in the number of minority-ethnic workers in the UK garment industry and a growth in informal and insecure work.

Informal production practices include small unregistered workplaces that pay no taxes or pay their workers , some of whom are homeworkers, cash in hand, along with more 'legitimate' businesses whose owners ignore labour laws to increase their profit margins. Interviews with trade unionists conducted for this research revealed that factories that may be registered and producing for large suppliers will often register the total amount paid to a worker with the Inland Revenue, and then divide that amount by the National Minimum Wage (NMW) to indicate the hours worked, whereas, in practice, an employee will have worked a much larger number of hours to earn that amount. Non-payment of the NMW and National Insurance to homeworkers is extremely widespread (Ellison 1999). Skimping on health and safety, working in inadequate buildings and non-presentation of formal working contracts are other practices that occur frequently. To understand the extent of informal working practices it must be remembered that a 'grey area' exists with many factories seeming to produce in ways perfectly in keeping with the formal sector, but employing their workers in conditions more in keeping with the informal sector.

Discussions with the AEKTA project, a worker's support project in Birmingham, revealed that while, officially, there has been a decline in clothing manufacture employment, it has seen no increase in unemployed workers, which suggests that at least some of these workers are being reabsorbed back into informal work. Estimates from trade unionists put the number of workers in unregistered companies at 65 to 75,000, which would make it only 20–25% smaller than the official figure for the industry.

This growth in unregistered informal companies would correspond with the growth in smaller-sized enterprises. Given the greatly increased downward pressures on manufacturers to produce at ever lower prices, owners have found other ways to cut manufacturing costs. A frequent complaint is that retailers refuse to pay a price for goods that would allow proper legal working conditions to continue. As smaller manufacturers, these firms have little power in subcontracting chains and are poorly placed to negotiate better terms for their products. The larger-scale factories that have now disappeared tended to be unionised, making it easier for workers to demand better pay and conditions.

The smaller enterprises that are growing in the industry today are much less likely to recognise unions and are much harder places in which to organise.

Minority-ethnic ownership of companies in the UK clothing industry has grown since the mid-1980s when recession and racism forced many Asian workers out of traditional manufacturing employment. The relatively low cost of business start-up in clothing manufacture means the industry was an attractive business opportunity (Basatia et al 1999). Thus, enclaves of minority-ethnic-owned businesses have developed, such as the knitting sector in Manchester, which represents the largest group of clothing employers in the region and is 98% Asian-owned (David Rigby Associates 2002).

The number of minority-ethnic workers in the industry is also growing, especially in smaller enterprises and those with minority-ethnic ownership. Informal working conditions are widely associated with a minority-ethnic workforce largely because immigrant workers have fewer labour market opportunities, being restricted in choice of job by language skills and legal status, and being prepared to undertake low status and low paid work traditionally done by women (International Labour Organisation 2000). The National Union of Knitwear, Footwear and Apparel Trades (KFAT) has found cases of immigrant workers being employed for as little £1 an hour in some factories and pay among homeworkers is notoriously low. A study by the National Group on Homeworkers (NGH) found that a much lower proportion of Asian homeworkers were paid the National Minimum Wage (NMW) than their white counterparts. Research by the AEKTA project (Basatia et al 1999) has also revealed pay as low as 83p an hour and an average pay rate of £1.90 an hour among Asian homeworkers.

The clothing manufacturing industry has traditionally been a female-dominated occupation, but this is changing rapidly. In 1995 over 70% of the national labour force was female, but Government labour market statistics from March 2002 showed that 41,100 men and 42,800 women were formally employed in clothing manufacture. This near equalisation is due to the much faster rate of decline of female jobs. Women had lost 8,800 jobs between March 2001 and March 2002, whereas men had lost 2,700 jobs over the same period of time. The largest number of redundancies in the garment industry are being made in female full-time jobs (Oxborrow 2002). Women are concentrated in the jobs which have been seen as low-skilled and labour-intensive such as sewing and pressing, and it is precisely these jobs that have tended to be lost overseas (CAPITB 2002). Men work mainly in management, cutting and warehousing, jobs which have been retained in the UK. The traditional UK workforce, which is white and female, is ageing and many have left the industry to retire or find more stable and better paid job opportunities elsewhere. Despite the large-scale redundancies, employers still report that, given the poor image of the garment industry, they have difficulty recruiting experienced staff and experience a serious skills shortage (Textile and Clothing Strategy Group 2002).

Official figures do not present the entire picture, however, as they say nothing about unregistered workers, many of whom have always been women. There is evidence that, here too, a significant shift is taking place. Although total numbers are unknown and homeworking may still comprise a large female labour force, interviews with homeworking support groups suggest that the numbers of women homeworking in the garment industry have been declining. There is also some evidence that the smaller, newer informal enterprises are employing a lower proportion of women than the larger and older factories. For instance, female employment only represents 30% of the total in the Manchester knitting factories even though knitting has traditionally been women's work (Parmar and Purdey 2000).

Overall, it is clear that current patterns of declining employment and industry changes have impacted hardest on women workers. It is they who have suffered most in terms of job losses, job insecurity and reductions in working hours. Women's paid employment has always been undervalued and seen as supplementary to the man's main 'breadwinner' wage. This perception of the industry as low-status, low-value women's work may well be why there has been so little public comment on the amount of job losses the industry has experienced in the last few years compared to industries such as car building or coal mining.

The inequality between male and female workers in the industry is most apparent in the difference in wage levels. In January 2002, men in the textiles and clothing sector earned an average of £303 per week while women earned an average of £209, only two-thirds the rate earned by men (Office of National Statistics 2002).

Trade union membership in the industry has plummeted from a high of around 32% to a current low of below 20%. The workforce of the larger old-style factories that had relatively high rates of membership has been disappearing, and trade unions have failed to make inroads into the new smaller companies in the sector employing ethnic-minority staff. Cultural and language differences between minority-ethnic workers and trade unions have played a role in this, and unions now recognise that they have to reach this segment of the workforce if they are to survive.

In 2001 the trade union KFAT had 5000 members in the North West region but by 2003 this had declined to only 2000. Indeed, the union, the only one in the UK to work specifically for the rights of workers in the clothing and textile industry, merged with another union, Community (previously the Iron and Steel Trades Confederation), in 2004.

Interviews with trade unionists conducted for this research also revealed the level of employer hostility they sometimes encounter from employers when they try to recruit new members. This can range from relatively well-off manufacturers assuring their workers that trade unions are not needed, to outright harassment and the dismissal of union activists. In unregistered companies, owners have often acted very aggressively against unionisation efforts and in these circumstances it is very difficult to hold any consistent union membership. The Transport and General Workers' Union (T&G) gained a membership of 500 informal sector workers in North London in the late 1990s but employers, despite legal protections being in place, orchestrated a campaign of lock-outs, intimidation and dismissals to resist such unionisation. Homeworkers have never been effectively organised by trade unions in the UK.

Declining trade union membership has left workers in the UK with even less of a collective voice in a time of extreme uncertainty. The growth of informal working practices has often left workers inadequately protected, making them extremely vulnerable to exploitation.

Research into the Clothing Industry in the UK

In the research reported here, three areas of the industry were chosen to develop an overview of different patterns of garment production in the UK They were, first, homeworkers who represent the most invisible area of production in the industry and the 'bottom end' of any subcontracting chain. Within this group of workers, two different sets of homeworkers, in Leeds and Rochdale, were studied. The women in Leeds were older white women who had previously worked in garment factories. The women in Rochdale were of Pakistani origin and had no previous experience of working outside the home. Asian women form a large proportion of homeworkers as language barriers, childcare responsibilities and recruitment from social networks restrict work opportunities outside the home (Basatia et al 1999).

The second point of entry was to look at knitting factories within an area of Manchester. This sub-sector highlights important characteristics that are predominant in the UK clothing industry more generally. The factories were Asian-owned, recruited mainly Asian workers, were very small in size and were situated within a relatively concentrated area. They tended to be characterised by informal working conditions, producing low-price goods for some high-street retailers and the market and wholesale trades. They tended to be at the bottom end of their subcontracting chains, struggling to compete in a market with very low margins. The third point of entry, in contrast, was to look at larger manufacturers who completed most of their production abroad while maintaining a small UK workforce. This group of companies were representative of those with more power in the clothing chain, and they tended to have closer relationships with their end users, add more value to their product and have a larger subcontracting chain beneath them.

Homeworkers

The homeworkers were involved in supply chains that were the most localised of all the groups researched. Not only were half working for sub-national markets, producing goods that would be sold to an end user in the same geographical area, but only two were producing goods that would be sold to major suppliers nationwide and they were the two producing non-fashion items. The subcontracting identified in this research tended not to include mainstream well-known high-street retailers (see Figure 6.1).

However, there was a marked contrast in the value of the goods produced by the women in Leeds and the women in Rochdale. Two of the Leeds women worked for bespoke (made-to-measure) tailors producing specialist and high-quality pieces for up-market hotels, casinos, public schools and the army. The clothing produced was made to the exact specifications demanded by the end user and would not be generally available. Meanwhile, the women in Rochdale produced, as is the norm for Rochdale clothing manufacture (Tyler 2001), either as part or all of their work, low-cost fashion clothing that was destined for small shops, markets or wholesalers. The Leeds chain could thus be said to be a high-value-added chain with homeworkers chosen for the high quality and specialist nature of their work, while the Rochdale chain relied on the ability of homeworkers to produce large amounts of clothing with general appeal at the lowest prices. The women producing in Rochdale worked in subcontracting chains where there were a large number of

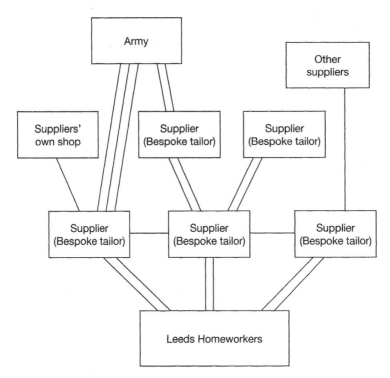

The number of lines indicates the ability of the subcontracted party to negotiate terms of production. One line indicates the most equal relationship between parties while three lines indicate the most hierarchical relationship.

Figure 6.1 Typical subcontracting chain of Leeds homeworker

them producing for each employer. However, the women in Leeds were the only homeworker producing for each employer, and they worked for three employers each. This means they wielded far more power within their chain, were more valued by their employers and worked in relationships that were far less hierarchical than the women in Rochdale (see Figure 6.2).

Dee's account, outlined in Box 6.1, illustrates the experiences of the group of homeworkers researched in Leeds. All of the women involved in the research were white and in their fifties or sixties. They had survived as clothing manufacturing homeworkers despite their reports of decline in the industry. Two had been actively sought out by their employers, one when she was considering retiring. This appreciation of

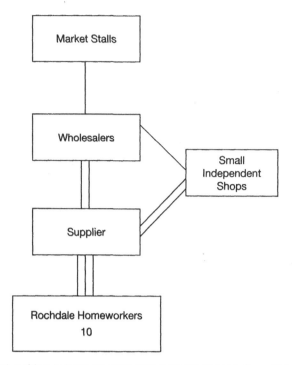

The number of lines indicates the ability of the subcontracted party to negotiate terms of production. One line indicates the most equal relationship between parties while three lines indicate the most hierarchical relationship. '10' indicates number of homeworkers.

Figure 6.2 Typical subcontracting chain of Rochdale homeworker

their work was cited as an important reason for job satisfaction and was a source of pride. All three said they liked working at home and would far prefer it to factory work:

> I have been happy homeworking. It has suited me. I'd really hate to work in a factory. They are like sweatshops.

Pay though was still very low for this group. All three said that the introduction of the NMW had not affected them because they already earned above this amount but calculating the price they were paid for completing each piece against the time it took them to do it revealed that two were actually beneath it. Not receiving work on a regular basis and not being informed of changes was cited as another problem:

Box 6.1 Homeworking in Leeds

Dee had worked as a garment maker since she left school at 16. At the time of the research, she was in her fifties and divorced. Her mother and grandmother were homeworkers making clothing, as were many women in that area. Dee followed changes in the clothing industry closely and was the only one of the Leeds interviewees to mention the MFA phase-out.

Dee worked for three suppliers, two of whom shared a building and rented off the third. All of the suppliers were bespoke tailors who worked on their own or only employed one helper. They made uniforms for the army and public schools, and one had a shop. She started working for her suppliers after they recommended her to each other. Dee's work was highly valued by her employers because she had multiple skills and could do all parts of garment production:

> He can't find anyone like me who can do the job lot. Do you know what I mean? If it came to it and it was just him I could go in and prepare them like I started off. Then I could bring them home and make them and I could go in and finish them and he could have his production whichever way he chooses to make them.

Dee saw homeworking as a career and was classified as self-employed for tax purposes, but this was the case even when she was working in the factory. She said the advantages are that she did not have to pay travel expenses or have a working wardrobe and she has time to do things like make dinner in the evenings.

The disadvantages are that she had no holiday pay, which meant she was without money for six weeks a year when the workshops closed. It's also a low paid job. Dee worked eleven hours a day and earned £10,800 in 2003, which works out less than the NMW. Not getting a pay rise with inflation meant she had lost 25% of her wage over the years:

> They were doing well last year so I got one, but before that I hadn't got one for four years... You are better off not pushing them or you are just pushing yourself out really because its not really affordable.

Despite drops in supply she had kept enough work by having three suppliers, but she found it difficult to balance all their demands. Overall though she said she would not like to do anything else:

> Well with me having done through-work I think you'd sell it more like a proper job wouldn't you? I mean it was full-time and it paid me a wage. And it's that you want to say really.

Summer it quietens off and outworkers are always first to be let go. They have got to keep the factory going. But it's just annoying when they don't let you know. They say they will sort something out and they don't get in touch with you. You wonder what is going on.

I think the more you do the more they expect. This particular one, she got to the stage she were fetching loads of work and expecting it back practically the next day. But you had to wait a fortnight for your money. You always got it. But she would never put herself out. She wanted her work regular but she didn't want to pay regular.... She thought if she couldn't come to your house because she had to go to the vet to fetch the dog you could wait. Instead of getting your wage Friday you could get it on Monday because she had to go to the vet's. But you need it over the weekend. There's just no choice.

This group of workers' experiences illustrate that although they were in a better position than many other homeworkers and all said that they enjoyed their work, they still faced many of the problems that are traditionally associated with homeworking. These included low pay, irregularity of work and pay, no holiday pay and being at the mercy of suppliers' timetables.

All the Rochdale homeworkers worked for suppliers with the same Pakistani origins as themselves. The three homeworkers interviewed received extremely low wages. Apart from Parveen, whose case is outlined in Box 6.2, the other two workers received respectively about £1.50 and £1.80 an hour. To work five hours a day for just £7.50 is to receive only around a third of the NMW. All of the Rochdale workers said they could not make enough to meet their basic needs. One of the workers asked her employer about the NMW. He told her, 'If you come to the factory I'll pay it.' But this has never happened, and the amount of work she gets has reduced. One of the homeworkers had received the same rate of pay for ten years.

All of the homeworkers said that they received work very irregularly and when work was available they often had to finish it in a rush. Work was also very seasonal in nature, as these quotes indicate:

They are very irregular. Sometimes they want it the next day on demand but then they won't turn up to collect it.

Box 6.2 Homeworking in Rochdale

Parveen was a separated mother of three. She had never worked in the garment industry outside her home although she would have liked to. She recently left the supplier she was making clothes for after four years to make punchbags for another supplier. She knew very little about her new supplier's company beyond the supplier's name. Her new supplier was another Asian woman. The punchbags were sold at Argos (a major retailer), which she knew because she saw them for sale in the catalogue.

Parveen was very happy to move to this new supplier as she was now paid regularly, although she still only earned around £3 per hour. Her old supplier used to pay her wage to her husband and at first she did not know how much she was earning. When she found out from other employees that she was getting less, she used to row with her employers but they told her 'if you don't like it don't work.' She said (through an interpreter):

> It was like that if you fall out with them. She only got paid irregularly so she left and he owes her money. She worked for pennies, and she still never got the money. The supplier was her husband's relative. She was really running around. Ringing up all the time to get the money.

Despite all the problems with homeworking, Parveen said the biggest problem is that all the homework is disappearing. All the local factories are importing from Pakistan and China. Because of this, she said, local employers cannot pay homeworkers more than they do or they would close. If her employer were to be prosecuted, he would not be able to pay the bills. He would close, and she would not get any work.

Ideally, Parveen would like to work in a factory because the pay is better but she cannot, as she cannot speak English.

No regular pattern. Sometimes when there is work she works the whole day. In September and October and January and February there is no work. It is different from working in a factory. The amount of work is on a supply-demand pattern. (Through an interpreter)

Unlike the Leeds homeworkers, the Rochdale homeworkers did not speak favourably about their work. Two of the women said that they would like to work in factories but could not because of childcare and/or language skills.

The biggest problem that the workers identified was a reduction in work. All had seen the amount of work, and the industry itself, decline in the last few years. The implicit or explicit threat that work could be taken abroad meant that these workers tolerated bad working conditions as other working options were limited by family responsibilities, language and fear of losing their jobs:

> It [home-based work] might just disappear in three to four years time. Her employer tells her they might have to close when they complain about something. (Through an interpreter)

The two groups of homeworkers in Leeds had issues in common despite significant differences in working conditions and attitudes to their work. Low pay, irregular work, declining levels of work, and isolation were problems experienced by both groups. Only one worker received pay equivalent to the NMW, though all had a right to it under the law. In addition to this, none of the homeworkers had ever received statuary sick pay, maternity pay or holiday pay. Only two homeworkers received wage slips for the hours they worked. All the women had bought their own sewing machines and bought equipment such as needles themselves. None of the women received any contribution from her employer towards the cost of heating and lighting while working from home. Finally, none of the homeworkers had received any advice on health and safety from their employer (see Gilbert 2002).

The research findings thus demonstrated the vulnerability of homeworkers situated as they are at the very bottom of any subcontracting chain. Even though circumstances can vary, the power of homeworkers to improve pay, conditions and the stability of their work is very limited.

Knitting factory workers

The knitting factory subcontracting chains explored in this research strove to remain competitive by producing at a low cost. The prices of knitted goods are falling in the UK market as discount retailers are gaining market share and many producers are sourcing abroad (David Rigby Associates 2002). The Manchester knitwear sector was found to

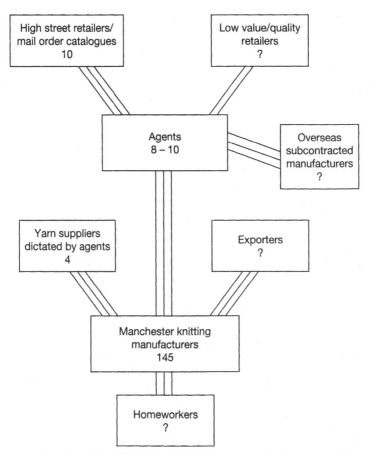

The number of lines indicates the ability of the subcontracted party to negotiate terms of production. One line indicates the most equal relationship between parties while three lines indicate the most hierarchical relationship.
Number indicates number of actors known at this point of the chain.
'?' indicates number of actors unknown.

Figure 6.3 Subcontracting map of the Manchester knitting factories

supply a range of end users from high-street retailers, mail order companies, exporters, wholesalers and markets to stores (see Figure 6.3).

The knitwear sector is vertically integrated and the only input to the factory is yarn that is spun, weaved, made up into the garment and packed within the factory. There are three or four actors in each chain, and each one exercises a high degree of power over the actor beneath. Retailers demand very low prices for goods, agents exercise complete control of the factories' supply and income and, if companies use homeworkers, extremely low wages are paid. There are estimated to be 145 knitting factories and around 8-10 agents in the Manchester area, so the individual power of factories to negotiate with agents is negligible. Moreover, in our research each factory was found to deal with just one agent so were very dependent on this relationship. There was no contact between retailers and the factories, and contact between agents and factories was often no more than a faxed demand to produce so many garments at a certain price in a certain time. Despite attempts headed by a local university to develop a partnership with the aim of fostering a cluster, strong competition existed between the factories. The factories were struggling to survive in a climate where they were expected to produce for a price that barely covered the cost of production.

The workers at the knitting factories often work in extremely poor conditions. The mills often house up to twenty small factories at a time and they are old, dirty and noisy. Windows are often boarded up, so there is not enough sunlight and poor ventilation. Lifts frequently do not work and workers have to climb many flights of stairs.

The workers are often employed on an informal basis with low wages and no job protection. Many of the workers are immigrants to the UK and so are restricted from finding work elsewhere. Most, but not all, of the women who work in the factories work in the sewing rooms. In one factory there was a cloth dividing men from women, and no contact between the sexes. The owner, a Pakistani woman, explained that the women preferred it this way.

As the knitting sector has faced such steep competition and price squeezes in recent years, factories now close down part of the year or lay off most workers during the quiet season. This was the case with all the knitting factories where interviews were conducted. All of them had also made workers redundant in the past year. Infrequency of work and job insecurity had made the knitting factories an unreliable and unattractive workplace for those with other job options (see Box 6.3).

Companies with outsourced production abroad

The companies that subcontracted some of their production abroad operated with far more negotiating power within their subcontracting chains, although two of the three had been forced to make substantial

Box 6.3   Working in knitwear in Manchester

Rasheed is from Pakistan and moved to the UK quite recently. He had worked at the knitting factory for three years. He tended to perform a variety of jobs at the company, depending on what needed to be done. The factory was in a very dilapidated building with little sunlight and was extremely noisy and dirty. The main complaint about the factory was that it was very cold in winter, as it had no heating and some broken windows. There was enough work at the factory for three months of the year and then three quiet months. For the last few years it has had to close six months of the year. During these periods Rasheed was unemployed, as other knitting factories in the area close too. In contrast, sometimes he had to work through the night at the factory though the women working there did not do these shifts. When there was work, the factory would stay open all day and night:

> There is no contract given to them, but it depends with the hours if the owner says there is 12 hours you have to work 12 hours. It depends how much work you do how much you get paid. You get paid for those hours you do. (Through an interpreter)

There was no higher rate of pay for overtime. Some people had come around from the KFAT union recently, but Rasheed felt that there was not much they could do for him:

> He isn't interested in joining the union, the reason being because there isn't much work as it is, so they really won't be able to do much. (Through an interpreter)

Rasheed said that people who can speak English do not want to work in the factories, as there is better paid work available elsewhere. Although Rasheed liked his job, he said the biggest problem is lack of work. Conditions cannot be improved without more work:

> He was saying that things were not going to get any better. He said that things are just going to get worse. Because there is no work available and what they need is work. If there is no work available how can conditions get any better? (Through an interpreter)

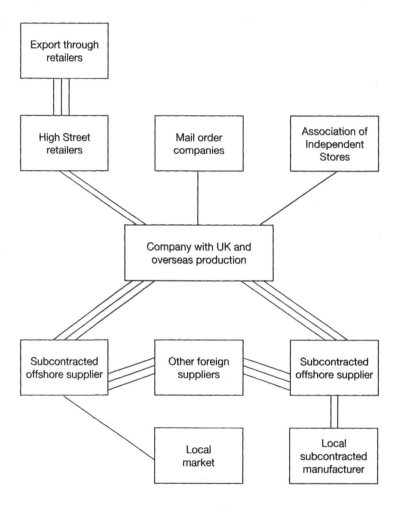

The number of lines indicates the ability of the subcontracted party to negotiate terms of production. One line indicates the most equal relationship between parties while three lines indicate the most hierarchical relationship.

Figure 6.4 Typical subcontracting chain of companies that produce both in the UK and abroad

redundancies and one of these companies was facing a very uncertain future. The vast majority of their production was for large retailers or specialist high-technology companies, and two worked directly with their end users. The contact between this group and the end consumer tended to be much more interactive with regular meetings and discussions. Each company was producing for a variety of end users (one had over 500 customers), making them less reliant on particular relationships and allowing them to spread risk.

The companies that subcontracted some of their production abroad were located much closer to the end user of their product than they were to the bottom links in their supply chain, with each producing in two different countries as well as running the manufacturing bases they operated in the UK. There was very little subcontracting in the UK, with only one company occasionally passing on work to another local factory.

All the foreign manufacturing operations were owned by local entrepreneurs, except for one that was part-owned by the UK company, and all produced for other suppliers as well as the interviewed company. This independence of the UK manufacturers and the foreign supplier was underlined by a lack of stability in the supply chain. All of the companies had changed factories and sourcing countries in the last few years and indicated they would do so again if another more competitive prospect was identified. A variety of reasons were given for choosing supply countries including quality, lead time, ease of communication and price, but a degree of randomness was evident in this choice. Suppliers were chosen on the basis of being invited to trade shows, mail shots and industry contact recommendations.

Workers in the factories of companies that produced both in the UK and abroad tended to be characteristic of the more traditional clothing manufacturing workforce (see Box 6.4). There was a very high percentage of female workers in all the companies interviewed, and these workers tended to be older. Pay and conditions were better than for homeworkers and knitting factory workers, and the required labour laws such as holiday pay, sick pay and so on were adhered to. All three of the factory owners complained about the legal requirements set for working conditions and pay, feeling that these regulations unfairly impinged on their competitiveness. It was clear that given the choice they would not follow these standards:

> We've got to be extremely careful if we employ someone and train them as a machinist, and then she gets pregnant for whatever reason, you know she wants a baby or she wants to go to the top of the council house list or whatever, that then costs us a lot of money personally. It's issues like that that we find unfair and obviously make us less competitive pricewise, because we have to factor all of these in.

Box 6.4 Working in a garment company with outsourced production abroad

Jean was a white woman in her fifties. She had worked for the company for 16 years and had been promoted to floor supervisor, a position of some responsibility. The company she was working for had just made massive redundancies and was reducing its UK workforce from 60 employees to around 12, as most of its production was moving abroad. All of the employees were women, apart from the owner and two men who worked in the cutting room.

The owner said that most of the workers were full-time, but Jean said the vast majority of women worked part-time. In contradiction to the owner, she also said that a lot of workers were let go when it was quiet, and were restarted later.

Jean obviously placed a great deal of importance on her work as supervisor and did not have any complaints to make about her working conditions. The redundancies had caused a great deal of tension in the factory, as some workers who had been there for sixteen years or more were made redundant over workers who had been there less than a year. Jean's sister was one of the women made redundant. Jean described the time as 'a nightmare.'

Many of the workers blamed the work from abroad for the job losses in the factory.

As work was sent out and came back through the factory, the women had to pack and check it:

> Perhaps the company has benefited [from work from abroad]. I don't know whether the workers would agree . . . They don't like passing any work from offshore . . . They class it as 'different work' but it is for the same company. And it's work at the end of the day, and you are going to get paid for it. You've not got a lot of choice.

The biggest problem for Jean was the insecurity of the company's future and the consequences of that for the UK workforce:

> I think we have just got to wait and see what is going to happen after. Whether things will pick up. Because there is not a lot of work coming in that is going to be done [here]. It's all small orders and things like that so at the moment I can't see what is going to happen . . . I don't feel secure because I don't know if I am going to be sat here this time next year.

Wages paid to workers ranged from around the minimum wage level to about £5.50 an hour. Two of the companies said the introduction of the NMW was an example of UK legislation that had a negative effect on their business, despite the fact that the NMW in Britain is still below any accepted 'living wage' level (Labour Behind the Label 2001). Workers involved in the subcontracting chains of companies that also produced abroad could be expected to have some of the better working conditions in the industry, although, comparatively, the work was still very low-waged.

The biggest problem for workers in this sector appeared to be redundancies and job insecurity. Nationally, this sector has lost the largest amount of jobs of all clothing production sectors. Two of the companies that carried out production both in the UK and abroad had made large numbers of redundancies and were looking to subcontract even more production to companies abroad. Paradoxically, the other company said moving work offshore had created better conditions for UK workers as more work was coming through the factory. The worker interviewed at this company said wages had also gone up.

Power and Knowledge Within Subcontracting Chains in the UK

The homeworkers interviewed had the poorest knowledge of the subcontracting chains they worked in. Only one knew how their supply chain worked beyond their employer. One Rochdale worker did not even know the name of the company she worked for, and most of the Rochdale workers had only a poor understanding of where the garments they produced were sold. This reflected the more distant and hierarchical relationships they had with their employers.

The factory workers could generally give a clearer picture of the subcontracting chains they worked in and could give details of at least some of the agents and retailers they produced for. Only two had any contact with actors in the supply chain outside their immediate factory. In one case, this was when work was dropped off and collected from the warehouse, and the other involved meetings where people from the retail outfit came to visit the factory for purposes of quality control. Workers in companies that produced both in the UK and abroad knew where garments were being produced as they handled them when the goods came back to their factory. However, they did not have a clear idea about whether these production facilities were owned by their employers or about the nature of relationships involved. None had had any contact with people from the offshore manufacturers, nor with actors further up the subcontracting chain. The lack of detailed knowledge held by the workers of the subcontracting chains of which they were part added to their sense of powerlessness.

Retailers clearly held the most powerful positions in the supply chains that embraced the Manchester knitting factories, and they then exploited this position to drive down the prices paid for garment production. Two of the factory owners said they preferred to work for the smaller, lower-quality retailers than the big chains because they paid better. Both had turned down work with high-street retailers because they would have lost money on the orders. As one explained:

> Cerrus [name changed] wanted us to do these jumpers for £6 each. But we would lose a pound on every one, so we did not want to do it. The agent said 'That is 10,000 pieces!' So I said 'Yes, but then we would lose £10 000.' He just didn't seem to get it. And they aren't going to change their margins no matter what.

Although agents undoubtedly exercised greater power within the supply chains than the factory owners, the agents were also susceptible to pressures from retailers. In a detailed government-funded case study (Travis and Greene 2000), one of the agents, who sourced from the factories in Manchester, was quoted as saying '[The retailer] has effectively driven us down on price as far as we can go—they can only squeeze us down so far.'

The low prices paid by retailers for garments and the very short time in which they expect manufacturers to deliver clearly had a negative effect on working conditions. As one factory owner explained:

> I've got two weeks, three weeks. So it is a very tight push. We don't mind the push, but when things go wrong it becomes a shove and that's what makes people ill.... It's very, very difficult to think clearly, and then you're pushing that on to the staff, which is wrong too. I'm not complaining about them [the supplier], but that's the situation, which is all they're interested in.

The factory owners were also vulnerable to changes in sourcing patterns by the retailers. When Littlewoods, a major UK retailer of knitwear, decided to stop sourcing in the area, some factories lost their main customer overnight. To balance the risk, some of the owners were trying to develop new markets such as export.

Without wanting to excuse those owners who would deliberately exploit low wages and poor working conditions to make a profit, it is clear that in some cases, owners may want to reinvest in staff and factories, or to legitimise their business, but the prices paid by retailers leave them no room to do this. While retailers profit from such conditions, they must also shoulder a great deal of the blame for the problems.

It is official government advice that companies must try to shorten their supply chain to become more profitable (Textile and Clothing Strategy Group 2002). For CMT factories, this means becoming designers and/or direct suppliers of clothing. The more progressive of the Manchester knitting factories owners were trying to develop ways to cut the agent from their chain by going direct to the retailer. As the only services agents provided were design and credit on yarn, this could work well. Manufacturers already carried most of the risk of production and paid for transport costs. However, the knitting factory owners were being frustrated by the lack of contact they could make with retailers. It is clear that the closer a company is to its end user, the more profitable it is likely to be and the more negotiating power it will have within the supply chain.

All of the manufacturers with production abroad reported that they kept a close relationship with the first tier of companies producing for them overseas, but they did this by directing orders via middlemen or agents, with no room for negotiating and dealing with the factory direct. These manufacturers thus had very little knowledge of how the supply chain operated outside the UK, particularly below the first tier. Two of the companies admitted that factories they outsourced to subcontracted work out when they were busy, but they had no contact with these 'second-tier' factories. It was a practice they said they 'frowned on,' but tolerated.

All three of the companies interviewed said they visited the factories producing for them abroad regularly, either at the start of each order or a few times a year. One employed roving quality controllers to visit these factories. There was an awareness of the sensitivity surrounding ethical sourcing, and all three were quick to say that they checked working conditions and health and safety in their overseas factories, but their commitment to the issue seemed shallow, as demonstrated in the following quote:

> They would always spend a day there at the most, three quarters of a day, so they would give the place a look over and measure a few garments to make sure that things look acceptable and then move on.

Ethical sourcing checks were done by quality control checkers or managers. There was no mention of external monitoring and, given the hazy knowledge of the subcontracting chain beneath the first point of contact, it seems unlikely that subcontracted units were checked. Moreover, it did seem that ethical sourcing had become a liturgy of 'acceptable' issues like child labour and broken needles policies. While these are important, a truly ethical sourcing policy would also have to include UK manufacturers examining their own sourcing practices: making sure workers were paid a fair wage and given a stable supply of work. Two company owners said they knew how much the workers in factories outside the UK were paid, but all issues of pay were left up to the local suppliers. These manufacturers had strict checks on quality control and other order specifications, which were backed up by visits, samples and roving quality checkers (see Popp et al 2000). One company even went so far as to send standard minutes to be spent on the production of each garment, but the same attention was not paid to foreign factory working conditions. Day-to-day running and working conditions were deemed to be solely the responsibility of local factory owners. The following conversation with an owner reflects this state of affairs:

CW: And you said you don't own them. Who owns them?
They are owned by the locals, similar to here, they are limited companies owned privately.
CW: So you don't have any stake in them at all?
No we don't.
CW: Just subcontract the work out.
Subcontract to them yes. Which is the same as global whatever it is.
CW: Globalisation?
Globalisation. Same thing. You just move on and that's the down side of globalisation. There is little responsibility for the indigenous population regarding the people or place of business there.

It is an important step that ethical sourcing issues have at least become an item on the clothing manufacturers' agenda, but any real movement on the issue seems a long way off in an industry that is so competitively driven by price. Retailers and large manufacturers must become willing to examine the effects of their own purchasing practices when considering ethical sourcing issues, if real progress is to be made.

Conclusion

Changes that occur in one part of the garment industry often have major repercussions for supply chains and workers' conditions. Decisions by large retailers, the most powerful actors in the UK garment industry, have far-reaching effects for all sectors of the industry, including those involved in supply chains which do not produce for them. Discount retailing, largely brought about by companies' increasing ability to source as cheaply as possible, has depressed pay rates and working conditions at both the high- and low-value ends of the market. Much of the infrastructure that was supporting homeworkers and the independent retail trade has disappeared causing a shortage of work and job insecurity in these subcontracting chains.

Changes in garment industry subcontracting chains have also affected the composition of the UK workforce. The research indicated that the movement of CMT jobs that were traditionally done by women to factories in countries with cheaper wage rates has resulted in a decline in female employment. The growth of smaller factories has resulted in an increase in minority-ethnic ownership and employment in the industry. The strategy of factory owners to survive by relying on informal working practices, such as non-registration of workers, non-payment of NMW and social security benefits and short-term employment, has increased.

Workers in the industry have very little voice. None of the workers interviewed were part of a trade union or said a trade union operated in their factory. In just one case, a union representative had recently visited after being invited in by the owner. In contrast, the homeworkers involved in the research were all in contact with homeworking support groups and valued this for finding work, getting help, filling in forms and acquiring information. The Asian workers found this support particularly useful, and the Rochdale group organised a regular meeting at which they would have a guest speaker. All of the workers felt this was very positive as it overcame their frustration and isolation.

Despite this positive experience, however, the research revealed that the garment industry has always had an image of being low-paid. There was a widespread acceptance amongst workers in each segment of the industry that things would get worse not better. The implicit or explicit threat that work would be lost or moved abroad meant workers would accept poor pay and conditions rather than risk losing their jobs. As one worker put it:

> Even if we got together it wouldn't help. I don't think unions or organising would help. Local factories are importing from other countries because they are cheaper. They can't pay us more because otherwise they would close.

A mood of pessimism about the future of UK garment production exists, therefore, among workers, trade unionists and manufacturers. The pessimism among manufacturers who produce in the UK prevents the development of long-term strategies and leaves employers little option but to reduce workers' conditions and wages in order to survive. For workers, a lack of confidence in their ability to improve their working conditions and an increased sense of competition with workers in other countries hinders any collective organising.

It is clear that workers producing in different garment subcontracting chains experience considerable variations in pay and conditions, which, in turn, are mediated through gender, ethnicity and legal status. Yet the intensification of global competition is squeezing workers' conditions across all sectors of the UK garment industry in familiar ways. A greater understanding of the operation of subcontracting chains, both in the UK and internationally, can highlight the nuances of a rapidly changing global industry and the commonalities of workers' experiences across space.

Note

1 In the 1950s the average size of manufacturing enterprises exceeded 60 employees, but today the average is under 25. An indication of the increased atomisation of the sector is given by a current 6% growth per annum in the number of clothing manufacturing businesses accompanied by the 11% per annum decline in sector employment (Oxborrow 2002).

7

The Impact of Full-Package Production on Mexico's Blue Jean Capital[1]

Lynda Yanz with Bob Jeffcott

Introduction

Until recently, Tehuacan, Mexico was considered a 'winner' in the global competition for garment industry investment and jobs. After the signing of the North American Free Trade Agreement (NAFTA) in 1994, the state of Puebla's second-largest city experienced dramatic growth in its jean manufacturing industry. By 2000, Tehuacan was competing with Torreon, Coahuila for the title 'blue jean capital' of the world. In that year, Tehuacan's garment industry produced 50 million garments a month, 40 million of which were for export. The value of garment exports was US$450 million, as compared to $200 million for the soda and bottled water industry, $250 million for the agricultural sector, and $100 million for the local commercial sector in the same region (Guerra 2002; Perez Cote 2001). At the end of that year, José Méndez Gómez, director of the Puebla delegation to the National Garment Industry Association claimed there was 'practically no unemployment' in Tehuacan and that people who didn't have jobs 'didn't want to work' (Ramírez Cueva 2001).

If the garment industry was a supposed winner under the new trade regime, Mexico's many indigenous peoples, including most of the inhabitants of the Tehuacan region, lost key historical rights.[2] The signing of NAFTA was accompanied by the dismantling of *campesinos'* constitutional right to communal land ownership and changes to water access rights that amounted to water privatisation. These changes had a dramatic impact on indigenous communities and cultures, which coincided

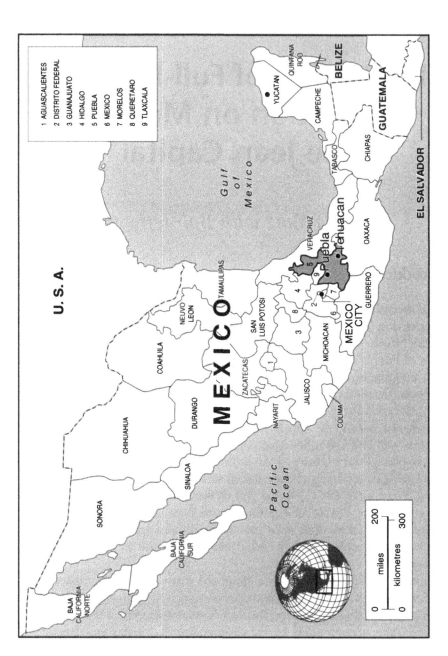

Figure 7.1 Map of Mexico showing Tehuacan

with the boom in *maquila* investment and, together with the lure of employment, encouraged the movement of indigenous youth from their traditional rural communities to urban wage labour in the *maquilas* of Tehuacan and its environs. Mirroring older colonial patterns, the young indigenous women and men in the region's garment factories work for a *maquila* industry controlled by a few elite families of Spanish origin. While migration of young people to the city is the dominant trend, in recent years some *maquila* assembly factories have also been set up in rural communities where wages and conditions are inferior to those in the city of Tehuacan.

In the chapter that follows, we present a detailed case study of the garment industry in Tehuacan (see Figure 7.1). After a brief overview of the industry's history in the region, we review the 'optimist' analysis of the Mexican apparel industry put forward by industry observer and academic Gary Gereffi. Gereffi and his colleagues have argued that increasing vertical integration, known in the apparel industry as 'full-package production,' is lending some long-term benefits to workers and their communities (Bair and Gereffi 2001; Gereffi and Martinez 1999). In contrast to Gereffi's findings, we present the results of research focused primarily on workers and their experience in the industry. This research, carried out by a local non-governmental organisation, demonstrates that working conditions and labour practices in the industry vary along the subcontracting chain. Furthermore, the *maquila* industry's extreme sensitivity to US economic trends impedes efforts to achieve even gradual improvements in labour standards.

The Tehuacan Garment Industry

Although Tehuacan's garment industry has a 30-year history, until the 1980s, manufacturers produced primarily for the domestic market.[3] The Mexican government's trade liberalisation policies of the mid-1980s paved the way for the signing of NAFTA in 1994, transforming the industry from production for domestic consumption to production for export, principally to the United States. The immediate post-NAFTA period saw the entry of major US retailers and brand merchandisers into Tehuacan's *maquilas*. The US jean manufacturer, Guess, was one of the first companies to shift production orders to *maquilas* in the Tehuacan region owned by local consortiums such as Grupo Navarra. In January 1997, *The Wall Street Journal* reported that Guess was planning to transfer the majority of its production to Mexico and other Latin

American countries, at least in part because of accusations by US unions and anti-sweatshop activists of sweatshop abuses in its Los Angeles contract factories, and increased vigilance by the US Department of Labor concerning its labour practices (National Interfaith Committee for Workers' Justice 1998). Other US brand merchandisers and retailers followed, including Levi Strauss, VF Corporation, Sara Lee, Farah, Calvin Klein, Tommy Hilfiger, Gap, Polo Ralph Lauren, and The Limited.

The greatest growth in investment in export production, employment, development of infrastructure, and automation took place in the five-year period after NAFTA came into effect. In response to the production demands of US brand merchandisers and retailers in the Tehuacan garment industry, some of the largest local manufacturers began to invest in new facilities and technology to upgrade their services from assembly to what is known as 'full-package production.' Full-package production refers to the co-ordination of a number of steps in the manufacturing process, from the acquisition or production of textiles and other inputs to assembly, finishing, packaging and sometimes distribution. While *maquiladora* plants assemble imported precut garments, manufacturers offering full-package services co-ordinate the various stages in the production process (Gereffi and Martinez 1999:2). Some US-based manufacturers, such as the Tarrant Apparel Group, also entered the scene, buying existing Mexican assembly plants, building new facilities, and creating full-package networks. Although most textiles and other inputs and machinery continued to be imported, some of the larger manufacturing consortiums that provide full-package services, such as Tarrant, invested in textile production facilities and/or purchased textiles from the domestic market. In addition, many of the larger consortiums established commercial laundries in the region to meet the style imperatives of North American jean consumers.

Is Full-Package Good for Mexico?

Gary Gereffi and Martha Martinez have challenged critics of NAFTA by suggesting that free trade is 'transforming the export sector in Mexico by substituting traditional maquila production with a new and more integrated form of export manufacturing,' i.e., full-package production (Gereffi and Martinez 1999:2). 'This full-package model,' they claim, 'provides better opportunities for development by accelerating technology transfer, creating high quality jobs with better wages, and providing opportunities for local entrepreneurs, correcting some of the exploitative characteristics of the maquila production system' (Gereffi and Martinez 1999:2). Gereffi and Martinez identify the development of full-package production networks as a significant step in industry upgrading, arguing that this trend has the potential to benefit workers and local economies alike. Unlike the *maquila* assembly model, which relies exclusively on foreign inputs and creates few backward linkages to local businesses, full-package production, according to Gereffi, could potentially support the development of local industries. He argues that investment in more capital-intensive textile and laundry plants could also signal more long-term investment in a particular region, as it has in some Asian countries (Gereffi and Martinez 1999:4).

Gereffi and Martinez came to these conclusions based on their research into the changes in investment, sourcing practices and the organisation of garment production in Torreon, Coahuila and the surrounding Laguna region. They observed that NAFTA and the devaluation of the Mexican peso encouraged the introduction of new manufacturing activities and the involvement of new types of buyers in the region. They suggested that US retailers and apparel merchandisers were beginning to see Mexico as an alternative to Asia as a source for their private labels and branded apparel. They also claimed that the presence in the region of major brand merchandisers with very detailed codes of conduct resulted in improved working conditions that 'often are better than those in similar factories in the United States' (Gereffi and Martinez 1999:6). While admitting that unions had been weakened by the federal government '[i]n tandem with the liberalisation of the economy' and that '[e]ffective representation and collective bargaining have almost disappeared,' they argued that workers used their mobility in a competitive labour market to negotiate small wage increases, productivity bonuses and non-monetary benefits (Gereffi and Martinez 1999:6).

In 2001, Gereffi, together with Jennifer Bair, carried out a second, more detailed study of jean production in the Laguna region, with less optimistic findings. While their study confirmed that there has been a 'significant shift' from *maquila* assembly production toward 'full-package networks characteristic of buyer-driven commodity chains,' the authors cautioned, 'the outcomes for local firms and workers are mixed' (Bair and Gereffi 2001:6). Specifically, their study showed that 'a significant portion of full-package orders in Torreon was being handled by a small number of first-tier manufacturers' (Bair and Gereffi 2001:10). Their study also found that the majority of the Mexican firms providing full-package services to foreign buyers were owned by 'family members related by blood or marriage' (Bair and Gereffi 2001:11). The significant amount of working capital needed to invest in and upgrade full-package facilities, as well as to purchase fabric, and the need for direct links to US clients clearly limited the possibilities for smaller local manufacturers to get into the full-package market. The study concluded:

> [T]he development of full-package networks in Torreon is primarily bene-fiting a wealthy domestic elite whose control over the local industry is being further strengthened by its exclusive access to the US customers placing orders in the region... While these orders are received by a few large, full-package manufacturers in Torreon, they are actually being filled by a burgeoning array of contractors and subcontractors organised into tiers of hierarchical networks controlled by the dominant firms in the cluster. (Bair and Gereffi 2001:11)

As a result, local firms were facing pressures from both US buyers and the first tier Mexican full-package manufacturers to 'reduce their production costs to a minimum in order to offer a competitive price' (Bair and Gereffi 2001:11).

Despite less than optimistic findings from this second study of garment manufacturing in the Torreon region, Bair and Gereffi continued to argue that working conditions were better in the larger full-package production facilities, and that the presence of the major brands using these facilities had improved working conditions (2001:13). Bair and Gereffi's assertion that the move to full-package production could lead to better wages and working conditions appears to be largely based on interviews with industry representatives and factory managers, observation during factory visits, and the assumption that voluntary codes of conduct and the threat of public exposure for links to sweatshop practices have been effective in eliminating the worst workers' rights abuses.

Tehuacan Research

In 2001, the Maquila Solidarity Network (MSN) and the Human and Labour Rights Commission of the Tehuacan Valley embarked on a joint research project on the impact of the restructuring of Tehuacan's garment industry on workers, indigenous communities and the environment. A key question examined was whether and to what degree the move to full-package production has benefited workers and communities in the Tehuacan region (Maquila Solidarity Network 2003a).

The Commission's access to workers and indigenous people was facilitated by its seven-year history of work in defence of the human rights of indigenous people of the Tehuacan Valley and surrounding communities. More recently, the Commission's work has focused on the labour rights of young, indigenous workers who come from those communities to work in the *maquilas*. As a local organisation, led and staffed by people from the area, the Commission was uniquely positioned to provide the , perspective of workers in the debate over full-package production and the general impact of the *maquila* industry.

The field research was conceived and carried out by two Commission staff between September 2001 and May 2002. The Commission undertook a review of official government documents, census and economic data as well as the main Tehuacan newspapers spanning a ten-year period from 1992–2002. The remainder of the data was collected through interviews. Five industry specialists working as supervisory and/or training personnel in the *maquilas* were interviewed in order to access an insider's technical perspective on changes in the industry. The majority of data was then drawn from 30 in-depth interviews with *maquila* workers.

Although some contacts were made on the street outside factories, all interviews were carried out in workers' homes, taped and later transcribed. Interviews were done in Spanish. Twenty women and ten men were interviewed ranging in age from fourteen to 42 years old. In selecting interviewees, the Commission looked for a balance of urban and rural experience: twelve interviewees lived in small towns in the Tehuacan Valley, eight were recent migrants to the city of Tehuacan, two had migrated to Tehuacan more than a decade ago, and eight were from the city originally. The eight workers from the city were *mestizo* and the other 22 were indigenous. With the exception of a fourteen-year-old worker, the Commission focused on workers with more experience in and knowledge of the *maquila* sector. The Commission also sought a balance of interviewees from different work environments. Accordingly, ten of the workers interviewed worked at large factories owned by the Grupo Navarra consortium, five workers were from the laundries, thirteen from small and medium factories and two from home-based workshops.

Interviews did not use a question guide but rather followed the narrative of the interviewee in her/his description of her/his labour experience. The interviewers added pointed questions pertaining to the industry, specifically about full-package production, flexibilisation and the module work system. In interviews with workers from the laundries, the interviewers also added questions about health and safety and environmental hazards specific to that work environment.

The home-based workers that were interviewed were from poor *colonias* of Tehuacan. In order to analyse the subcontracting system at this level, the two investigators observed the transportation of products to and from the *maquilas* by vans and by foot. The families involved in the research were contacted through an extended family member that had been involved with the Commission. In addition to a more formal interview with the heads of the families (women in both cases), the interviewers talked informally with the rest of the family members while they were sewing and finishing garments. They also observed the number of pieces in the home and the rate of production. During the course of identifying possible interviewees, personnel from a small *maquila* in the neighbourhood threatened one family with loss of work if they participated in the interview.

The research project coincided with the downturn in the US economy, which negatively affected Tehuacan's garment industry and garment workers. Since the completion of the research, the impacts of the recession in the US have continued to be felt in Mexico. Media reports of *maquilas* fleeing Mexico to lower-wage countries such as China, Honduras and Haiti are common. Reports suggesting that long-time brand-name customers (such as Wrangler and Lee) might pull out of Tehuacan are also beginning to appear. Current wage levels in Tehuacan and other garment centres, which, according to this and other reports, do not meet workers' basic needs, are often blamed for the flight of jobs.

Industry Hierarchy

According to the Commission, it is virtually impossible to determine the exact number of garment facilities at any given time due to the precarious and often underground nature of light manufacturing. Companies often close down in one neighbourhood and reopen under a new name in another; others operate illegally in peoples' homes (Mendez 2001). In 2000, according to official figures of both the Tehuacan municipal government and the national garment industry association, there were approximately 300 garment factories in Tehuacan (Ayuntamiento Municipal de Tehuacan 2002; Ramirez Cueva 2001). In fact, these figures greatly underestimated the actual number of manufacturing facilities since they only took into account those registered with the *Secretaría de Hacienda* or affiliated with the business associations. According to the Commission, there were as many as 700 garment-manufacturing facilities in the Tehuacan region in 2000, including large factories owned by the major local and US consortiums, medium-sized independent factories, and small clandestine sewing workshops employing fifteen to twenty people. The Commission's estimate did not include home-based facilities.

Similar to Gereffi and Bair's observation of first-tier companies in the Laguna region (see above), export manufacturing in Tehuacan has been dominated by a few large consortiums owned by Mexican families and a prominent Los Angeles-based apparel manufacturing family. These large consortiums, including Grupo Navarra, AZT International, Tarrant Apparel Group (TAG-MEX), and Mazara, had direct relationships with US retailers and brand merchandisers to whom they offered half-package to full-package services (see Figure 7.2).

At the time of writing, the Fernandez family controlled two major consortiums in the region. Grupo Navarra was principally Mexican-owned by the Fernandez family, although US investors reportedly also owned shares in the company. Grupo Navarra had seven assembly plants and two laundries located in Tehuacan. Before the US economic downturn, its factories produced up to 150,000 jeans a week for Sun Apparel, Guess, Wrangler, Gap, Levi Strauss, Polo Ralph Lauren, Tommy Hilfiger and Sara Lee, among others. Its seven assembly plants did cutting and assembly, and its two laundries—Cualquier Lavado and Lavapant de Tehuacan—did laundering, sand-blasting, pressing, inspection, labelling, sizing, final inspection, and packaging. Mazara was another Tehuacan-based consortium owned by a sub-clan of the Fernandez family. The company owned three factories in the Tehuacan region—Confecciones Mazara, Confecciones Rotterdam, and Industrias Cerraquin. Confecciones Mazara was formerly part of the Grupo Navarra consortium. Mazara has produced for Gap, VF Corporation, and Guess, as well as other brands. Mazara did not own laundry facilities, but offered laundry services by using facilities owned by other companies.

The Tarrant Apparel Group is a Los Angeles-based consortium. At the time of the Commission's research, Tarrant owned eight assembly and laundry facilities in the Tehuacan area and one textile mill in Puebla. It was also involved in the construction of a twill mill, garment-processing facility, and distribution centre in Tlaxcala. The founder, current Chairman of the Board, former CEO and major stockholder in Tarrant is Gerard Guez. Another major stockholder in, and former president of, Tarrant is Kamel Nacif, known in Mexico as 'the Denim King.'[4]

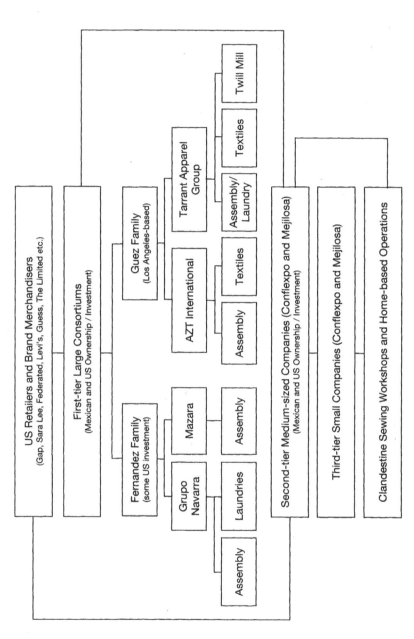

Figure 7.2 The Tehuacan garment industry hierarchy

Tarrant has produced for Gap, Tommy Hilfiger, Express (The Limited), Levi's, Federated Department Stores, Charming Shoppes, The Wet Seal, and other specialty retailers, discount retail chains, and brand merchandisers. AZT International is a related but separate Los Angeles-based consortium also associated with the Guez family and Nacif.[5] The company produced textiles in their mill in Parras, Coahuila, assembled in Tehuacan, and did laundering, labelling, finishing, and packaging in Panzacola in the neighbouring state of Tlaxcala. The company's assembly plants were the most high-tech in the region, utilising some automated sewing processes. AZT International has produced for Calvin Klein, Express, Gap, and Tommy Hilfiger, as well as other brands. At times, the company also subcontracted some work to medium-sized companies.

The second tier of Tehuacan's apparel manufacturing hierarchy includes medium-sized companies, most of which are independent factories owned by local capital. They produce both for US and domestic brands and retailers, and sometimes do subcontract work for the large consortiums. Employment of under-aged workers is more common in these factories than in those owned by the large consortiums. At the time of this research, some of the medium-sized companies, such as Confexpo and Majilosa, were collaborating to provide half-package services, and were planning in the future to be able to provide these services independently. Some second-tier firms subcontract parts of production, such as attaching buttons, buttonhole sewing, and thread removal, to clandestine workshops and home-based facilities.

At the lowest rung in Tehuacan's garment industry hierarchy are the small and clandestine companies. The sewing workshops and home-based facilities owned by these companies are dispersed throughout the city and the surrounding region, and are often illegal, underground operations. Some change their location constantly to avoid detection. Many of these companies manufacture products for the domestic market, including shirts, pants and jeans. Some also do subcontract work for the medium-sized companies. One of the main goals of small companies is to obtain orders to assemble products for export. Home-based garment assembly is common in Tehuacan and surrounding communities. The home-based workshops are concentrated in both established and informal *colonias* surrounding the city, such as Mexico, Las Palmas, Benito Juarez, Granjas de Oriente, La Paz and Juquilita. Homework also takes place in rural communities in the region like Ajalpan, Altepexi, Chilac and Zinacatepec.

Home-based work is done on a subcontract basis for small and medium-sized manufacturers in order to lower the cost of production.

This work is usually labour-intensive, detailed and repetitive, such as removing threads from apparel, sewing buttonholes, and/or attaching buttons. While the vast majority of homeworkers are women, other family members, including young children, often participate in production. Women homeworkers also often hire other women to work in their homes. The use of under-aged workers, as young as ten or eleven years old, is most prevalent in these small assembly factories and home-based sewing workshops. Pay is strictly by the piece, the work is intensive for long hours at a time, and workers usually do not receive the legal minimum wage. The 'employer' receives on average 50 *centavos* per piece, and the workers she employs might earn between 25 and 30 *centavos*, or eight *pesos* per bundle. In an eight-hour period, one person might be able to complete 300 pieces, and would receive about 32 pesos (Cdn$4.99 or US$3.20).[6]

Although the federal labour law covers homework, the legal requirements are generally not respected. At the time of the Commission's research, the National Commission for Minimum Salaries had established the minimum wage for homeworkers at 50.85 pesos (Cdn$7.97 or US$5.11) for an eight-hour day. Since payment is strictly by the piece, homeworkers' right to the minimum wage was being systematically violated. Employers are also required by law to register their use of homeworkers, and provide employment contracts describing conditions of employment. According to the Commission, neither of these requirements is being enforced. Nor do homeworkers receive overtime or statutory holiday pay. Finally, workers in home-based workshops do not receive Social Security (health care) benefits.

Impact of US Downturn: 2004 Update

The economic downturn in the US and the fallout from the September 11 attacks drastically deflated Tehuacan's nascent export-dependent industry. The much-anticipated end of apparel quotas worldwide in 2005 has further eroded Mexico's position in the industry since the downturn (see Chapter 9 for more about the impact of the MFA phase-out). Industry sources in the Tehuacan region counted a loss of 20,000 garment jobs, falling from 35,000 to 15,000, between October 2000 and December 2001, as a result of the partial or total closure of 150 plants out of the official total of 300 (Perez Cote 2001). The Commission estimates that the actual job loss in the Tehuacan garment industry as a result of the downturn in the US economy and the September 11 attacks was

25,000 jobs, a reduction in the workforce from 70,000 in October 2000 to 45,000 in May 2002. The downward trend continued in 2003, in which at least 8000 *maquila* jobs were lost. Many of the factories that remained open moved to a '3X4' workweek, in which employees worked for three days each week followed by four days off, cutting production by up to 50%.

A number of companies reduced workers' wages and eliminated production bonuses and various other benefits.[7] In addition, the largest apparel consortium, the Tarrant Apparel Group, closed its eight plants in the region between January 2003 and February 2004, eliminating 6000 jobs. Tarrant, as the largest regional producer, had previously set the wage standard in Tehuacan paying between 750 and 1500 pesos a week (Cdn$117.56-235.13 or US$75.41-150.83). With the closing of Tarrant, Grupo Navarra set the new standard at an average range of 450 to 1100 pesos a week (Cdn$51.82-115.17 or US$40.07-89.05).[8] This standard put downward pressure on the lower rungs of the industry. Wages fell to a range of 350 to 500 pesos per week (Cdn$40.31-57.59 or US$31.17-49.52)[9] for workers in medium and small *maquilas*. Workers at the lower end of the wage scale in the larger factories received fewer benefits and were often denied Mexican social security (IMSS) by their employers (Barrios and Hernandez 2004). Although exact figures are unavailable for 2004, given the structure of the industry, one can assume that wages in rural *maquilas* were adversely affected as well.

The US economic downturn not only had an impact on garment workers and manufacturers, it also negatively impacted other sectors providing inputs or services to the industry and its workers, such as food vendors, store owners and employees, and people who rent living quarters to migrant workers. These developments have called into question overly optimistic forecasts of long-term economic growth and full employment anticipated by the proponents of free trade and the full-package model.

The Tehuacan *Maquila* Workforce and Their Working Conditions

Eighty percent of workers in maquila garment factories in the Tehuacan region are young indigenous people who have migrated from small rural communities in surrounding mountainous regions, as well as other regions in the states of Puebla, Veracruz and Oaxaca (Ramirez Cueva 2001). The social consequences of the migration of rural indigenous youth into the urban *maquila* workforce include a high percentage of single mothers, violence in the home, and a high incidence of venereal disease, including AIDS (State Centre for Municipal Development 1999).[10] At the time of writing, the gender make-up of the workforce was split evenly between male and female in the assembly plants.[11] In the laundries, all the production workers were male.[12] The majority of the workers were between the ages of fifteen and 30.[13] Some of the smaller assembly facilities employed workers as young as eleven years old.[14]

Before the US economic downturn, worker turnover was a major problem for maquila owners. According to workers interviewed by the Commission, moving from factory to factory was one of the only strategies available to them to negotiate improvements in their incomes during the years they were gaining experience and skills. Some workers reported having worked in more than ten factories, from large modern factories of the Grupo Navarra to underground sewing workshops. They attributed the high turnover rate to the low wages in the region and the resistance of employers to raising salaries. According to workers, employers in the neighbouring community of Ajalpan developed and were circulating a blacklist with names of workers who were fired or voluntarily quit their jobs. The list was reportedly used to weed out job applicants who commonly moved from factory to factory, and to discourage that practice among other workers.

The most common complaint of workers interviewed by the Commission was that wages were too low and did not meet their basic needs. At the time of interview, most garment assembly workers in the Tehuacan region were making between 350 and 1000 pesos a week (Cdn$54.86-156.75 or US$35.19-100.55). A few more skilled workers, however, such as the *encuartadores* who sew the inseam of jeans, made up to 1100 pesos a week (Cdn$172.42 or US$110.60), and some workers in the laundries made as much as 1200 pesos a week (Cdn$188.10 or US$120.64). In jean assembly factories in the neighbouring rural regions of the Sierra Negra, the municipality of Vicente Guerrero, and Santa Catarina Otzolotepec, assembly workers were being paid between 250 and 300 pesos a week (Cdn$39.18-47.02 or US$25.13-30.17). These wage rates represented a significant decline since the period prior to the US economic downturn and September 11. In 2000, wages for assembly workers were between 750 and 1500 pesos a week (Cdn$94.35-188.71 or US$65.26-130.52).[15] It appears that *maquila* owners used the economic downturn, with the resulting employment insecurity due to layoffs and plant closures, as an opportunity to reduce labour costs.

The Commission calculated the cost of basic necessities by asking interviewees to add up their basic food, rent and transportation costs for the week. Based on these interviews, a worker with three children living in Tehuacan required 970 pesos per week (Cdn$152.05 or US$97.53). This figure did not include expenses for children to attend school, nor the costs of shoes or clothing. In order to meet the families' basic needs, workers reported they often relied on the informal economic activity (e.g., selling food outside the house) of other family members to supplement their *maquila* income.

In addition to wages that did not meet basic needs, workers had several other common complaints reviewed briefly here and elaborated in more detail below. Workers reported compulsory overtime, unpaid overtime, fines and deductions for mistakes or tardiness, lack of seniority rights, and the failure of a number of factories to register workers with the government social security programme (IMSS). In terms of harassment, workers reported verbal abuse and humiliation by supervisors, especially racial insults and discrimination toward indigenous workers, discrimination against pregnant women, including pregnancy testing, and sexual harassment and abuse. The absence of freedom of association was identified as a major problem. Workers complained that independent unions were prohibited and protection contracts rather than collective agreements applied at their workplaces. Several workers reported the use of child labour in small and home-based facilities. In addition to these broad categories of complaints, laundry workers frequently complained about their exposure to toxic chemicals.

Workers interviewed by the Commission reported that they were often forced to work extra hours without any compensation in order to complete their daily production quota. It was common for workers to work ten to twelve hours a day without earning the legally required overtime pay. This practice appeared to be one of the most common violations of the Federal Labour Law. Workers also reported that when they did meet production quotas in the time allotted, they often found that the quota was then increased, with no increase in pay. While the practice of requiring workers to work overnight (*veladas*) in addition to their regular shift during heavy production periods has been less common since the US economic downturn, it continued to occur in some factories.

A number of workers interviewed commented on the problem of child labour and exploitation of under-age workers. One worker commented:

There are a lot of minors of nine, ten, eleven years of age working in the maquilas when they should be in school. This happens because unfortunately their parents aren't paid enough money to provide the basic necessities for their children. I have two nephews, one thirteen and the other fifteen years old; the second one is working at [name of a small factory], and many times he leaves work at 8 at night.[16]

A worker employed at that factory confirmed there were many under-age workers. She described the problem in the following words:

The kids are able to do the work, but they aren't paid the same wages as adults who do the same work. Some minors are paid 150–200 pesos a week, nothing more. There are children from nine to fourteen years old working in the factory. Some are manual labourers, but some work on the machines. When Social Security arrives, the supervisors hide them…

A minor working at the factory described his situation in these words:

I'm fourteen years old and I work in a *maquila*. There are other children working in the factory, some fifteen, some my age. They put me to work attaching rivets on the sides of the jeans. I have to make my quota. They pay me 400 pesos, but one month ago they told me they're going to give me a raise. They said they were going to give me 500 pesos, but they haven't given me anything. Adults that complete the same production as I do are paid 600 pesos, but because I'm a kid, they don't pay me the same.

Of the money I'm paid, I give 300 pesos to my family, and I keep 100. I don't like working in the *maquila*, but I have to do it because I have younger brothers. My father doesn't make enough money to provide for them, and I have to help pay the household expenses. After work I don't do anything because I'm tired and I just want to go to bed. I eat dinner and I lay down. In the morning, I have to go to work. On Sundays, I stay at home and wash my clothes. When I have time, I play soccer with my friends. I would like to go to school like most children do.

Gender-based discrimination was apparent in the hiring practices of employers. The increase in male participation in the garment sector in Tehuacan appeared to reflect the introduction and growth of jean laundries, exclusively employing male production workers, as well as the growing number of men who are compelled to seek sewing jobs out of economic necessity. At the same time, subcontract sewing workshops and home-based facilities employed only women, girls and some under-age males. Compulsory pregnancy testing was another form of gender-based discrimination in at least some of Tehuacan's *maquilas*.

In addition to gender discrimination in hiring and promotions, discriminatory treatment of indigenous workers appeared to be a common problem in Tehuacan garment factories. An indigenous woman worker described the racial insults indigenous workers are often subjected to by employers and supervisors in the following words:

> I've seen a lot of examples of verbal abuse and insults of workers by bosses and supervisors. They say things like: 'you're a donkey,' 'get moving stupid,' 'get to work you lazy Indians.'

Sexual harassment was also identified as a problem by some of the workers interviewed. A former employee of a factory owned by one of the large consortiums described her experience with sexual harassment in the following words:

> Three weeks after I started working at [name of factory], the line supervisor began to harass me. He would say that if I would go with him to a hotel, he would give me an exit pass and I wouldn't have to complete my daily production quota. Because I wouldn't accept his offer, one day he told me to leave at 9:00 at night. The next day, I was called in to the boss's office, but instead of sanctioning the supervisor, they told me that I was fired, with no explanation why. A lot of women have to endure these pressures, because they need to work to feed their children. For a woman, it's really difficult to work with these kinds of abuses.

Health and safety was a serious concern for workers, especially with the introduction of additional industrial processes as the industry moved to the full-package production model. Possibly the most serious health and safety hazard in Tehuacan's garment industry was exposure to toxic chemicals in jean laundries. Interviews with laundry workers indicated that little attention was being paid to health and safety training, preventive measures or provision of personal protective equipment. At the time of the research, a tragic accident at the jean laundry Cualquier Lavador, owned by Grupo Navarra, resulted in the deaths of two workers. On November 27, 2002, Martin Bernardino Hernandez and Raul Sanchez Vazquez died, and Daniel Leon Gonzalez and Arelio Valencia were injured while doing maintenance work on a cistern in which residual water from the laundering process was recycled. The cause of death was reported to be exposure to chemical fumes. The company was fined 250,000 pesos for violation of the *Ley de Protección Civil* (civil protection law) (Osorio 2002; Teran Soto 2002).

Interviews with laundry workers carried out months before the accident at Caulquier Lavador indicated that while they were not fully aware of potential health problems, laundry workers were very concerned about the possible negative impact of exposure to chemicals on their health.

A worker at the laundry of Exportadora Famian, owned by the Tarrant Apparel Group, who was more aware of the chemicals being used than most of his fellow workers, explained his concerns:

> Every day, we're exposed to toxic substances—fumes from caustic soda and chlorine, contact with enzymes, detergents, peroxide, oxalic acid, sodium bisulphate. Every day, we breathe and are in physical contact with these substances, because the company no longer gives out facemasks because they say we're exposed to gases, not to large particles. I have a sewer's facemask and some plastic gloves, and when they break, the company's not going to want to replace them.
>
> All of my workmates have respiratory problems and sore throats. The most extreme case I've seen was a guy whose nasal passages were injured by the bi-sulphuric gases, and they bled for two weeks. They treated it as an illness, and not as a work accident.
>
> I've been sick for five months. I have a fungus on my hands from contact with enzymes they use in the laundering process. I went to see the company doctor, and he told me I had a skin fungus and that I should go to the social security clinic. Even though I have social security, I've had to pay, and I've lost a lot of work time recovering.

Most injuries reported by assembly workers were related to production pressure. Workers interviewed reported that accidents were common in assembly factories, particularly needle punctures of fingers. They attributed the prevalence of accidents to the pace of production, long hours of work, and pressure to complete production quotas.

The failure of employers to register workers with the Social Security government health care programme (IMSS) was also a common problem in many medium-sized and small factories. A woman employed by a medium-sized company explained how her employer's failure to register her with IMSS caused her to lose her maternity leave benefits:

> I'm in my ninth month of pregnancy, but I'm continuing to work even though it's uncomfortable and difficult for me to meet my quota. I continue to work because this month the company gave me Social Security. They didn't want to register me, but I kept insisting and pressuring the boss. But now that I have Social Security, they're not going to pay me during maternity leave, which is for 40 days before and 40 days after the birth, because I wasn't able to catch up with the required contributions

after being registered so recently. Neither IMSS nor the employer is taking responsibility for my rights. I need the money because having a baby is expensive, and if the baby is ill, it's going to be very difficult.

Although the Commission found many violations, workers also reported some improvements in the *maquila* industry in the Tehuacan region. Prior to the US economic downturn, some of the larger factories owned by the major consortiums, particularly those that produced for well-known US brands, saw improvements in health and safety and some of the more egregious violations, namely the elimination of child labour and pregnancy testing. There was a general increase in the use of personal protective equipment, such as masks, goggles, hard hats, gloves, and work boots. The large consortiums introduced fire protection equipment, increased workspace, improved ventilation and lighting, and reduced noise levels. Some training on health and safety and codes of conduct was introduced in larger factories, but usually for supervisors rather than production workers. Vaqueros Navarra, a factory previously known for its compulsory pregnancy testing of employees, eliminated the practice. Some factories installed daycare centres, though in the case of factories owned by Grupo Navarra, the facilities had insufficient space to meet the demand. However, some benefits introduced during the period of business expansion were subsequently eliminated due to the US economic downturn.

It is worth noting that the introduction of full-package services did not directly lead to improvements in labour standards in the industry, but rather introduced some brand-sensitive companies to the region. According to the Commission, the improvements seemed to be isolated to those factories producing for brands that were particularly sensitive to allegations of sweatshop practices. The elimination of some of the most visible abuses–those most disturbing to North American consumers– such as forced pregnancy testing and child labour could be at least partially attributed to the increased vigilance of US brand-name buyers and their factory monitoring programmes. Other less consumer-sensitive issues, however, such as poverty wages, unreasonable production quotas or targets, forced and unpaid overtime, and violations of freedom of association continued in the large factories, as they did in smaller facilities. In addition, the number of brand-sensitive companies using full-package services is relatively small compared to the retailers, discount retail chains, manufacturers and brand merchandisers that are less vulnerable to anti-sweatshop campaigns and less rigorous in monitoring factory conditions.

In fact, one could argue that the introduction of new actors to the *maquila* industry creates as many problems as it solves. Retailers and brand merchandisers and the development of full-package networks to meet their ever-changing fashion demands led to work intensification and exposure to, as well as discharge of, a variety of toxic substances used in the laundering of designer jeans. Although there were some improvements in quantifiable standards such as health and safety practices, some of these changes could be attributed to the requirements of more modern machinery, such as improved ventilation and space between machines.[17] Finally, interviews with workers producing for major brands in factories that had improved indicated that most workers continued to be unaware of codes of conduct or how they might be used to promote improvements in working conditions and labour practices.

Freedom of Association and the Right to Bargain Collectively

One key reason that garment workers in Tehuacan and other garment centres in Mexico have been unable to negotiate significant improvements in wages and working conditions, even in periods of economic growth and full employment, is the absence of independent, democratic unions or free collective bargaining. In the Tehuacan region, the dominant labour federation is the FROC-CROC, an 'official' union linked to Mexico's historical ruling party, the PRI, and to the current Puebla state government. The CROC (Revolutionary Confederation of Workers and Campesinos) is known for its negotiation of 'protection contracts' with employers without the knowledge or participation of workers, and often before any workers have been hired.[18] In exchange for this protection, unelected leaders of the official unions receive automatic dues deductions from their 'members" pay cheques.

A worker interviewed by the Commission who worked at Grupo Navarra's Lavapant laundry described his experience with the CROC in the following words:

> When I started working at Lavapant, I was given an employment contract that had a lot of pages with very small lettering that was almost illegible. I started to read it closely and especially a section that referred to a union, which said that the moment I signed the contract, I would automatically become a member of the FROC-CROC. Because I was taking time to read

the document, the management person started pressuring me, saying, 'If you want to work here, you better sign immediately because there's a line up of people outside who want jobs.' I needed the job, so I signed the contract. Six months later, I still haven't met my supposed union representative, and I'm beginning to wonder if this person really exists.

A second worker employed by Exportadora Famian, owned by the Tarrant Apparel Group, explained why workers are reluctant to report worker rights violations to representatives of the CROC in the following words:

> When you have problems and you bring them to the union, the leaders and representatives always side with the boss, never with the worker. Who does a union serve when they tell you to bring your problems to the boss? Now they are saying there isn't any work, there isn't any money, the boss hasn't received any orders. Just the same, you need money for medicine, and they tell you, no, they can't help you. But, they keep deducting union dues from your salary.

At the time of writing, the only independent union with a signed collective agreement in the maquilas in the state of Puebla was SITEMEX (Independent Union of Mex Mode Workers) in the Mex Mode, formerly Kuk Dong, factory in Atlixco. The factory produces sweatshirts for Nike and Reebok, including for the collegiate market. The workers began to organise in 2000 around low wages and rotten food served at the company cafeteria. After a three-day work stoppage in January 2001 ended in a violent attack on the workers by state police, workers received a great deal of international support from students, unions and NGOs in North America and Europe. The strength of the workers' organisation combined with an international campaign led to the signing of a collective agreement in September 2001.[19]

The achievement of an independent union and signed collective agreement at Kuk Dong/Mex Mode raised hopes that this successful campaign combining local organising with international pressure on brands and institutional buyers could serve as a model for other organising efforts in the region. However, similar campaigns that followed, including those at Tarrant and Matamoros Garment, were much less successful. The particular combination of circumstances and convergence of forces at Kuk Dong/Mex Mode made this case somewhat unique. These included the presence and vulnerability of major brand buyers sourcing from the factory, pressure from US university buyers motivated by student action, the involvement of multiple monitoring organisations and multi-stakeholder initiatives, the technical sophistication of the manufacturing facility and the commitment of the owner to keep the factory open, the fact that the full impact of the economic downturn had not yet been felt, and the strength of the workers' movement on the ground. This combination of factors is the exception rather than the rule in Tehuacan's garment industry. It is also worth noting that the breakthrough at Kuk Dong/Mex Mode increased the resolve of manufacturers, 'official' unions and government officials in the region to resist further efforts to achieve independent unions.[20]

The only other example of an independent union in Puebla's garment industry was the SUTIC (Garment Industry Workers Union), which represented sewers in Puebla, Tecamachalco and Tehuacan in the 1980s. In 1989, the leader of the union, Gumaro Amaro, was assassinated, reportedly by gunmen hired by the state governor of the time, Mariano Piña Olaya. The SUTIC did not survive the ensuing government repression (Reyes 1997).

Currently, the unions that do exist in Tehuacan's garment export industry, most of which are affiliated with the FROC-CROC, though some are also with the CROM (Regional Confederation of Mexican Workers) or the CTM (Confederation of Mexican Workers), can be found in many of the larger factories and some of the medium-sized plants. The very existence of the official unions in these factories, and their institutional ties with the state government and the Conciliation and Arbitration Boards that review and grant union registrations, prevent authentic worker organising and collective bargaining from taking place. This does not, however, mean that workers do not engage in job actions when they feel their rights are violated. Over the past six years, there have been a number of wildcat strikes and other spontaneous job actions in garment factories in the Tehuacan area. Most of these actions were in response to the failure of the employer to pay legally required wages or overtime pay, unjust dismissals, and/or the failure to provide legally required severance pay at the time of dismissal.

As a result of the US economic downturn and the resulting plant closures and layoffs in the Tehuacan garment industry, there was a significant increase in reports of labour rights violations in 2001. There was also a change in the kinds of violations reported and the manner in which workers responded to them. The vast majority of the claims before the Local Conciliation Board in Tehuacan and the Labour Tribunal in Puebla were for unjust dismissal and/or for failure to pay the legally required severance pay. It is worth noting that of the 591 claims brought before the Labour Tribunal, only 53 resulted in the payment of a financial penalty by the employer (Junta Local de Conciliación de Tehuacan 2002). These violations were not restricted to small firms facing threats to their economic survival. In a case before the local Conciliation Board in October 2001, workers charged their employer, Cualquier Lavado, owned by Grupo Navarra, with unjust dismissals, forced labour (compulsory unpaid overtime), failure to pay severance, and physical and verbal abuse by the employer and his armed guards (El Sol 2001). While the lack of independent unions, collective agreements or effective state institutions to adjudicate worker rights violations has been a major obstacle to workers' efforts to improve their conditions in periods of economic growth, these same factors have resulted in the further victimisation of workers during periods of economic downturn.

Postscript

Since the completion of our joint research project, both MSN and the Commission have been involved in campaigns to defend the rights of garment workers employed by the Tarrant Apparel Group in Tehuacan and the neighbouring community of Ajalpan. These efforts have raised a number of issues about the possibilities and limits of brand-focused campaigns and code monitoring programmes in influencing the labour practices of the largest consortiums that have established full-package networks in the Tehuacan region.

In June 2003, workers at Tarrant's Ajalpan factory staged a work stoppage to protest against treatment by management and the company's failure to provide legally required benefits. As often happens in Mexico's *maquilas*, the leaders of the work stoppage were fired and the protest was quickly transformed into a fight for an independent union. An international campaign was launched to pressure major brands sourcing from the factory to ensure that their supplier respected the workers' rights. While the campaign was successful in winning the intervention of some of the brand buyers, Levi Strauss and The Limited in particular, it was not successful in winning the reinstatement of the fired workers or the employer's acceptance of the union. Tarrant refused to co-operate with a Levi's-initiated factory audit, despite the fact that this resulted in the loss of future Levi's orders. While it did eventually co-operate with an audit for a more important client, The Limited, Tarrant refused to take corrective action based on the finding of that audit, even though this refusal threatened its relationship with a major customer.

Significantly, other buyers sourcing from the factory were less willing to put pressure on their supplier.

As typically happens when independent union movements emerge in Mexico, the Local Conciliation and Arbitration Board dismissed the workers' application for union registration on technicalities in October 2003. The Board's decision to dismiss the application was made despite the overwhelming support for the union in the factory and significant pressure from a major brand, Levi's. Levi's sent letters to the state authorities, urging them to ensure that the workers' petition for union registration was dealt with in accordance with the Federal Labour Law. Tarrant has since closed the factory, and although there is some speculation it will reopen the plant under a new name, it is unlikely that any supporters of the independent union will be rehired.

Meanwhile, in Tehuacan, the Commission had been providing advice and support to hundreds of workers from other Tarrant-owned plants who were illegally dismissed without proper severance pay. In one Tarrant factory, workers formed a worker coalition and staged a protest through the streets of the city. While the Commission was not successful in winning the reinstatement of the fired workers, it did help them gain most, if not all, of their legally required severance pay. As of February 2004, Tarrant had closed six factories in the Tehuacan region, laying off approximately 6000 workers. At the time of writing, it remained unclear whether some or all of the Guez family's investments in the Tehuacan region would be shifted to other countries, or whether ownership of those facilities would be shifted to another Guez brother or back to their Mexican partner, Kamel Nacif. If Nacif does take over ownership of some or all the factories, Tarrant could resume sourcing from those factories when the economic situation improves.

Supporting workers' efforts to form independent organisations and to demand respect for their rights continues to be a perilous pursuit in Mexico. On December 30, 2003, Commission Coordinator, Martin Barrios, was physically assaulted by an unknown person. He has also received death threats. A media campaign against the Commission continues.

Conclusion

NAFTA and the trade liberalisation policies of the Mexican government contributed to the dramatic growth and restructuring of Tehuacan's garment industry, and encouraged the entry of new players—US brand merchandisers and retailers—and the development of full-package networks. While the move toward full-package production increased the amount and types of Mexican inputs in the production process, as well as the elements of the supply chain located in the Tehuacan area, it also encouraged the further concentration of wealth and power in the hands of a few members of the local elite and a California-based family. While the growth and restructuring of the garment industry in Tehuacan created needed jobs for young indigenous workers, it also caused negative social, cultural, economic and environmental consequences for indigenous communities. Changes in land ownership and agricultural policies, also as a result of trade liberalisation policies, coincided with the growth of the garment export industry, encouraging the migration of indigenous youth to wage labour in the garment export industry and contributing to the deterioration of indigenous agricultural communities.

Involvement of major apparel brand merchandisers and retailers in the Tehuacan garment industry created new leverage points to challenge the most flagrant worker rights violations. As a result, there appear to have been some improvements in working conditions and labour practices, at least partially as a result of code monitoring programmes and increased public awareness of these issues in Mexico, the US and Canada. However, those limited improvements appear to have taken place only in larger facilities owned by a few US and Mexican consortiums with direct relationships with high-profile brands. Some of these improvements were rolled back as a result of the economic downturn. Manufacturers producing for less brand-sensitive retailers appeared not to be under the same pressure to improve working conditions and labour practices. As well, areas of improvement in the larger factories tended to be on 'hot button' issues that are of most concern to North American consumers, such as child labour and forced pregnancy testing, and on quantifiable problems easily identified through factory monitoring, including some health and safety practices. Other persistent, but less dramatic problems, such as low wages, high production quotas, long hours of work and compulsory and often unpaid overtime, gender and race discrimination, and denial of freedom of association, have not received the same level of attention.

New workplace problems and community issues have emerged as a result of the entry of brands and retailers into the Tehuacan garment industry, and the move to full-package and modular production. These issues include work intensification and the health and safety problems associated with it, and health and environmental problems associated with chemicals used in laundry processes. Some garment assembly has shifted to home-based facilities in Tehuacan and small workshops and factories in rural areas where wages are even lower and conditions are practically unregulated.

The US economic downturn has had serious negative impacts on workers and communities in the Tehuacan region. The downturn, and resulting layoffs and plant closures, is easily utilised by employers to cutback on labour costs and weaken worker rights. The recent factory closures and worker layoffs at Tarrant factories in Tehuacan and Ajalpan call into question whether the move to full-package production guarantees more stable investment and employment and opens up the possibility for successful worker organising. Furthermore, Tarrant's willingness to close factories in the face of a union organising drive and its refusal to cooperate with major customers' efforts to enforce their codes of conduct point to the limitations of voluntary codes and brand campaigns in influencing the practices of large consortiums owned by wealthy and powerful Mexican and US investors, particularly in periods of economic downturn. The events at Tarrant and other factories have shown that the successful worker organising effort at Mex Mode/Kuk Dong is not easily replicable.

Despite new leverage points created by the presence of major brands in Tehuacan's garment industry, the alliance of apparel manufacturers, 'official' unions and state institutions continues to be a major barrier to workers' efforts to organise to improve their wages and working conditions. The real and perceived threat of intensified competition with Asian, Central American and Caribbean countries in the wake of the quota phase-out in 2005 makes efforts to improve conditions in Tehuacan's garment *maquilas* even more difficult. New strategies and alliances will need to be developed to increase pressure on companies and government to respect workers' rights, and create new incentives for apparel suppliers to comply with those rights.

Notes

1 This article is based on *Tehuacan: Blue Jeans, Blue Waters and Worker Rights* jointly authored by the Maquila Solidarity Network and The Human and Labour Rights Commission of the Tehuacan Valley (2003). Martin Barrios and Rodrigo Santiago Hernandez of the Commission carried out the research in Tehuacan. We would also like to acknowledge the contributions of Marion Traub-Werner and Georgia Marman; both assisted with editing this article.

2 In fact the city was originally called Tehuacan, City of Indians (Paredes 1910).

3 Interview by the Human and Labour Rights Commission of the Tehuacan Valley with Javier López, director of the local delegation of the Cámara Nacional de la Industria de la Transformación, Canacintra, January 29, 2002.

4 When Tarrant purchased the Puebla denim mill from Nacif in March 1999 for 2,000,000 shares of Tarrant common stock and US$22 million in cash, it also announced the appointment of Nacif as president of Tarrant México. It is worth noting that Nacif is not only one of the wealthiest men in Mexico, but is also well connected with important political leaders at the state and national level.

5 The parent company, Azteca Production International, is 50% owned by Paul Guez, and the company's CEO is Hubert Guez. Both are brothers of Gerard Guez. Paul was the founder of Sasson Jeans, the original designer jeans. Gerard was the head of Sasson's Los Angeles office until he left in 1998 to found the Tarrant Apparel Group.

6 Unless otherwise noted, all currencies converted using the rates quoted for September 1, 2002 at *www.xe.com/ict/*.

7 Between July and December 2001, local media, including *El Mundo de Tehuacan* and *El Sol de Tehuacan* published numerous articles quoting employers, workers and municipal government representatives concerning these developments.

8 Converted using rate quoted for January 1, 2004 at *www.xe.com/ict/*.

9 Converted using rate quoted for January 1, 2004 at *www.xe.com/ict/*.

10 An interesting finding in this study is that *maquila* employers have resisted participation in governmental campaigns on reproductive health, claiming that they would cut into production time.

11 This reflects national trends in the garment industry where in 1999, 48% of garment workers in border regions and 54% of garment workers in non-border regions were women.

12 Based on official statistics on the Economically Active Population in Tehuacan contained in the Municipal Economic Census, Municipality of Tehuacan 1999–2002.

13 Based on worker interviews carried by the Commission.

14 Information on child labour is from the Commission's interviews with parents, relatives and co-workers of minors employed in the *maquilas*.

15 Converted using rate quoted for January 1, 2000 at *www.xe.com/ict/*.

16 Mexican labour law permits the employment of fourteen- and fifteen-year-olds with their parents' written permission. However, those young workers cannot work more than eight hours a day.

17 MSN interview with Homero Fuentes, Director, Commission for the Verification of Codes of Conduct (COVERCO), concerning improvements in health and safety practices and production in full-package facilities in Central America, July 2002, San Pedro Sula, Honduras.

18   These employer/union agreements are called 'protection contracts' because they usually mirror, and sometimes undercut, legal requirements, and provide protection to employers against the threat of authentic worker organising and collective bargaining. For more information, see RMALC and Alexander and La Botz (2003)

19   For more information on the Kuk Dong struggle visit the Maquila Solidarity Network website: *www.maquilasolidarity.org*.

20   During a separate independent union organising drive in the hospital sector, the governor of the state of Puebla pledged to deny all union applications for legal status, promising '[N]ot one more union until the end of my term.' *La Jornada de Oriente*, June 10, 2004: p. 3, author's translation.

8

Defending Workers' Rights in Subcontracted Workplaces

Rohini Hensman

Introduction: The Challenges Facing Garment Workers in Defending Their Rights

Previous chapters have shown that the subcontracting arrangement in itself creates an environment in which workers' rights can, and are, easily violated. The basic problem is that, unlike the usual employment relationship, where the firm that profits from the workers' labour employs them directly and is therefore accessible to collective bargaining, here the firm that profits most from the labour of the workers—the retailer or brand-name company—may be located thousands of miles away, and is certainly not accessible to collective bargaining. Chapters 5, 6 and 7 have described some of the resulting problems for workers on the ground, and this chapter explores what can be done to solve them. In this chapter I look at the research material specifically from the point of view of workers attempting to struggle for their rights and those who wish to support them, in order to identify which strategies are effective and point to others that might work in future.

If we are thinking of strategies to defend workers' rights in this predominantly hostile environment, it helps to classify workers into three main groups, depending both on their place in the corporate chain and on their employment status. At the top of the chain are regular (or permanent) workers in large-scale first-tier units, either within or outside export processing or free trade zones (EPZs/FTZs). Next come irregular workers—those who have temporary or seasonal contracts—working for large, medium or small-scale (first to third-tier) units. Finally we have informal workers—those who have no contracts at all—working predominantly in small or medium (second- or third-tier) units or as homeworkers. The problems are somewhat different for each of these groups, and, more importantly, strategies that work for one group may not help another. Breaking down the subcontracting chain in this way will help us to arrive at some conclusions about ongoing efforts to defend workers' rights in these chains, and how they can be made more effective or supplemented by complementary demands and struggles.

Regular Workers in Large-Scale Units

A small minority of workers in this sector are unionised and have been able to demand and win decent employment conditions. One example is Pirin Tex Production, one of the case studies examined in our research in Bulgaria, employing nearly 2000 workers with decent pay, good working conditions, and compliance with labour law (Women Working Worldwide 2003a:212). However, most workers even in this sector are not so lucky. Many work in EPZs/FTZs where there is either an implicit or explicit deal between government and employers that unions will be banned. For example, until the mid-1990s in Sri Lanka, despite the fact that in theory the normal labour laws of the country applied to EPZs/FTZs, in practice, workers attempting to fight for their employment rights could expect to be victimised severely, losing their jobs, and at one time possibly even their lives. Fortunately, this disjunction between legal and actual rights has provided a space for workers to fight for the right to organise. After the People's Alliance government came to power in 1994, the new Labour Minister announced that workers in EPZs/FTZs would henceforth have the right to unionise, but faced with a backlash from employers, who threatened to pull out their capital en masse, the government retreated, substituting Workers' Councils for unions. Some of these Workers' Councils were employer-dominated, but some workforces took the opportunity to establish genuinely representative workers' organisations, and on 30 June 1996, the Joint Association of Workers and Workers' Councils of the Free Trade Zones of Sri Lanka was launched. In December 1999, parliament passed an amendment to the Industrial Disputes Act, making it mandatory for an employer to recognise a union if 40% or more workers in the unit belonged to it, and following this, on 23 January 2000, the Free Trade Zone Union of Sri Lanka was formed. On 9 March 2003 it merged with the Industrial, Transport and General Workers' Union to form the Free Trade Zone and General Services Employees' Union (FTZGSEU). The struggle to get recognition for unions from employers still continued, but the fact that the FTZGSEU subsequently succeeded in getting some of its units recognised in the EPZs/FTZs of Sri Lanka is a tribute to the militancy and persistence of the workers, unions and support groups involved (Clean Clothes Campaign 2001; Transnational Information Exchange-Asia 2004).

Where workers in this sector have not been able to organise, they face serious problems at work. Work during peak periods is so intensive that they are not even allowed to go to the toilet when they need to do so, and, combined with long hours, health hazards and virtually no off-days or holidays, this often leads to exhaustion and illness. Overtime and night work are usually compulsory, in the sense that workers who refuse are liable to be fired, or workers are locked in and not allowed to leave until they have completed the allotted quota. Sexual harassment at work is supplemented by the even worse risk of sexual assault on the way home from work late at night. Stories like that of Zarina and Delowara in Chapter 5 are all too common; whether in Bangladesh, Pakistan, Sri Lanka or elsewhere, women workers returning home from work late at night seem to be considered fair game for molestation, rape and murder by a section of the male population, and the lack of any protection from such attacks should be seen as yet another occupational hazard they face. The horrifying scale of rape-murders in Ciudad Juarez in Mexico led to a massive movement of protest which could provide a model for women workers elsewhere (Wright 2001; see also the example of the Mex Mode factory campaign in Mexico, Chapter 7).

Consumer campaigns to force retailers and brand-name companies to adopt Codes of Conduct have been used most successfully on behalf of this section of workers.[1] To begin with, these are the most visible suppliers, those that are monitored and inspected if their buyer has a code of conduct, and these are the workers who are also most visible, being on the payroll of those suppliers. Even without a struggle, there have been cases where an improvement was observable; for example, the researchers in Bangladesh found that: 'the wage rate is significantly higher in the factories that have direct contact with the buyers than that in the factories which have no direct contact with the buyers... mainly due to the fact that in the former types of factories, buyers monitor the compliance of their codes of conduct' (Women Working Worldwide 2003a:45). It is significant that the only successful struggle to unionise garment workers in Bombay has been in the two factories of Excel, with 250 and 150 workers respectively, which supply Van Heusen, and are monitored for compliance with that company's code. Moreover, where struggles to unionise such factories have been resisted by employers, solidarity campaigns by consumer organisations, trade unions and others in support of the workers can be successful. For example, the workers of the Gina Form Bra factory in Bangkok won their struggle for union recognition and reinstatement of sacked activists on 9 July 2003, after more than two years of struggle, with support and solidarity action from activist groups in Thailand and many other countries (Maquila Solidarity Network 2003b).

Solidarity action along the subcontracting chain also has most potential for formal workers in large-scale units at the top of the chain. We had one example in our project where information provided by workers in supplier factories in the South helped the workers in a retail unit in the North to get their union recognised. The latter was Pinault-Printemps-Redoute's (PPR's) Brylane Kingsize division in Indianapolis, USA, where health and safety problems as well as the right to organise were major concerns for the workers. In the course of a sustained campaign to organise themselves as a unit of the textile and garment union UNITE, they collected information from suppliers of Brylane in Thailand, the Philippines, Romania, Saipan, Indonesia, Pakistan and India, including Patel Hosiery in Bombay, demonstrating serious violations of workers' human rights. Combining this with information about health and safety hazards in the Indianapolis unit itself, and injuries suffered by the women who worked there, UNITE printed a 'Sweatshop Catalogue,' in the style of a normal glossy catalogue, which was distributed to customers and shareholders at a shareholders' meeting in Paris. PPR was sufficiently embarrassed to recognise the union and sign an agreement with it in 2002. As a letter dated February 10, 2002, sent by UNITE to those who had helped them by supplying information, announced: 'On January 29th, Brylane finally agreed to respect the workers' demands and recognised the union. For the first time in the history of Brylane's Indiana facilities, a committee of elected worker representatives will sit across a negotiating table from Brylane management to bargain a union contract.'

The letter continues: 'In addition, in response to UNITE's public reports of the working conditions that workers endure at Brylane's sourcing factories... Brylane announced that it has adopted a Code of Conduct which incorporates the core labor rights of the International Labor Organisation (ILO) and is intended to cover workers at all suppliers of PPR's Redcats division, of which it is a part (UNITE, 2002, personal communication).' When a Marathi translation of this letter was read out to a group of women workers, including women from Patel Hosiery who had supplied information which was used in the campaign, some of the women broke out into spontaneous applause, pleased and excited that they had helped workers in a distant corner of the world to get organised, and hopeful that they too might benefit from the agreement. However, UNITE went on to express dissatisfaction with the code as well as the suggested monitoring mechanisms, and hoped that in future they would be able to sign a proper agreement with Brylane to strengthen the code and stipulate effective mechanisms for monitoring it and ensuring its implementation. Such an agreement would be a step forward from a Code of Conduct, because unlike codes, which consumer campaigners have so far been unable to get courts to recognise as being legally binding, *a collective agreement is legally binding*, and potentially a more powerful means of defending the rights of subcontracted workers. Efforts by unions to negotiate International Framework Agreements with retailers and brand-name companies which cover their suppliers could result in better protection for workers at the top of the subcontracting chain, including those in developing countries (see Miller 2004).

The relationship between buyers and suppliers is important for this section of workers. Overwork and compulsory overtime during peak periods are likely to be endemic problems so long as the current 'just-in-time' policies as well as the fashion-driven nature of the industry continue. Our research work found that women garment workers in Tamil Nadu, observing that orders that they formerly had to complete in a month now had to be completed in a fortnight, asked, 'What's the great hurry? Will people in Europe and America have to go without clothes if we take a month to complete the order?' (P. Swaminathan, personal communication, 2002). Obviously there is no such pressing reason for these tight schedules, and since garments—unlike fruit, flowers and vegetables—are not perishable, employers cannot claim this as an excuse either. Consumer campaigners have questioned the policy of saving on storage at the cost of violating the rights of workers who produce the garments sold by retailers, and also the craze for constantly changing fashions. Indeed, consumers would benefit from a change of this kind by having a more predictable offer available for sale in the shops. If retailers were to order some core items on a more long-term basis, this would certainly reduce the irregularity of work for women producing the garments and allow consumers to be able to rely on finding the basics on store shelves.

Even if the retailer is sincere about upholding workers' rights in first-tier suppliers, however, it may not always be possible to do so. In one of the factories located on an industrial estate in Parel in Bombay, Go Go International (owned by one of the biggest quota holders in India), women workers participating in our research sought the help of office employees, who managed to get a list of all the company's buyers. They found they were supplying an astonishing sixty-four companies in twelve different countries! Two of these companies had codes of conduct, but they did not help very much, as we found subsequently. Anjali, a militant worker who had previously been dismissed from Contessa Knitwear for trying to form a union, was one of those who in January 1999 signed up around 110 workers in this factory to a union too.[2] The response from the employer was swift: on January 24th the workers were locked out on the plea that there were no orders, although it was clear to the workers (and confirmed by the company's website, which boasted that orders were always plentiful!) that this was not the real reason. By the end of March, a settlement had been arrived at, whereby the workers resigned 'voluntarily,' but would be paid their dues and retrenchment compensation, with contributions to the union being deducted from each. Anjali passed on the information she had gathered about the codes of conduct to the union organiser, who used it to negotiate a higher level of compensation for the dismissed workers than they would have got otherwise. But there was no way of getting them reinstated or getting the union recognised *despite* having these codes of conduct in place (Women Working Worldwide 2003a:93,100).

What this example suggests, and other experience confirms, is that codes of conduct work best in those first-tier factories where the bulk of production is for just one or two buyers who have codes and are serious about implementing them. Where there are numerous buyers, as in the case of Go Go, it is easy for the supplier to let go of one or two buyers who insist on workers' rights, because so many more are still left. On the other hand, if several buyers have codes of conduct, that, too, becomes a problem. The multiplicity of codes is confusing for the supplier as well as the workers, and the need for each of them to be monitored separately is a huge waste of resources. In such cases, it would be much better if all the retailers had a common code and a common monitor who could be relied upon to ensure that workers' rights are being respected. SA8000[3] could potentially play such a role, but reports suggest that the organisations on the ground that are accredited to do the auditing often have no expertise in detecting violations of workers' rights (Shepherd 2001).

Since this is the section of workers that is most visible and best covered by labour law, codes of conduct and consumer campaigners can help to ensure that the law is implemented in countries where the government supports their rights. However, where the government itself is at fault (for example in EPZs/FTZs, or under repressive regimes), pressure needs to be put on the government to enact and implement legislation protecting basic workers' rights. Linking these rights to multi-lateral trade agreements of the WTO would be one way of putting pressure on such governments. There has been a great deal of contention about the proposal for a so-called social clause which would do this, mainly on the grounds that it could be used as a protectionist measure by developed countries. This is indeed a danger, but if a social clause can be formulated in such a way that it cannot be misused in such a manner, it would be a useful tool for workers in first-tier supplier factories. If it can be argued that violating the basic human rights of workers is an un-acceptable way for employers and governments to compete in world trade and must be ruled out, this would make it much easier for workers in these factories to organise and bargain collectively to improve their wages and working conditions. The involvement of the government is also essential to ensure that children who are taken out of employment in supplier factories are fed and educated rather than being pushed into even worse forms of exploitation (see, for example, Cleveland 2003; Fields 2003; Hensman 2001; Munck 2002).

Irregular Workers in Large, Medium and Small Units

These workers are mostly concentrated at middle levels of the supply chain, but a few may be found at the top. The very fact that they have short-term temporary or seasonal contracts makes it much harder for them to organise. If they have been taken on because extra workers are needed to complete an order on time, and they know that once the order is completed they will have to leave, they may not be motivated to organise, knowing that although they are here today, they will be gone tomorrow. This is likely to be the case even in first-tier supplier factories, which can justify the use of these workers to inspectors or monitors by referring to the exigencies of production. A more insidious situation arises when the same workers actually work for the same employer over a long period, but are given short-term contracts either in order to avoid paying them anything during slack periods, and/or to discour-age them from organising. There were such cases identified by our research in Pakistan, and it has also been reported quite frequently in other research and campaigns in Korea. Here permanent women workers were dismissed and then rehired as temporary workers (also called 'dispatch workers') through employment agencies (Women Working Worldwide 2000:7,19). It is almost certain that if such workers made any attempt to organise, their contracts would not be renewed. This situation is even more likely to arise in second- or third-tier sub-contractors and in the medium to small units, including sweatshops, that contract in work from the larger factories, none of which would be inspected by the government or monitored by the buyers.

In addition to the problems suffered by the first category of workers (such as overwork, long hours and compulsory overtime during peak periods, health hazards and sexual harassment), this group is also likely to suffer from low pay, long stretches of no pay at all during slack periods, and no social security or welfare benefits. Indeed, there are employers who quite deliberately employ workers on short-term contracts when they could easily be made permanent. This allows the employers to avoid paying social security contributions or maternity benefits, and to enforce long hours and low pay without much fear of organised protest from workers, since these workers can be terminated once their contract is over without any reason being given. Razia, working in the finishing department of Daisy Knitwear in Lahore and still temporary after three years, is an example. In our research work, she told us that women in her factory did not get maternity leave, and there was no union in the factory because if any worker tried to form a union, he or she would be terminated without any reason. She started work at 7.00 a.m. and finished at 10.00 or 11.00 p.m., earning only R1200/-per month, less than half the minimum wage. She said: 'We work long hours without getting overtime. My male supervisor misbehaves and harasses me by passing unwelcome remarks... If I refuse to do overtime, the supervisor and managers ask me to leave' (Women Working Worldwide 2003a:133).

Defending the rights of these workers, most of whom come lower in the supply chain than the first category of workers, is more difficult. The very insecurity of their employment makes them reluctant to risk losing their meagre income by making demands for improvements, either individually or collectively. Thus the first requirement is to strengthen their legal rights. This is a challenge to the trade union movement, at both local and international levels. Trade unions should ensure that wherever workers are employed on short-term contracts, these workers must be paid at the same wage rates as permanent workers doing work of equal value, and get the same facilities (like paid off-days, holidays and leave) and benefits (including contributions to health insurance and retirement benefits) on a pro rata basis. They should also ensure that minimum wage rates are adequate to allow the worker to pay her social security contributions as well as survive. In India, employers have waged a media war to entrench the idea that job stability of any kind makes a firm inflexible, which in turn makes it uncompetitive in the world economy. Their primary argument for using irregular workers, therefore, is flexibility. Workers' organisations need to respond by arguing that if what they really want is flexibility, they should be willing to pay 'flexible' workers at least as much as they pay permanent workers, if not more, instead of paying them a fraction of what they would have to pay the regular staff.

If temporary, casual, seasonal and part-time workers cost the employer the same as full-time regular workers, there would be much less temptation for employers to substitute the former for the latter. Ensuring that they receive wages, paid leave and benefits at the correct rate will mean that trade unions/workers' representatives will have to translate piece-rates for each operation into daily and monthly wage rates so that the calculation can be made, since most of these workers are piece-rated. Additionally, trade unions should press for labour law to stipulate that workers who could be employed on a permanent basis because the job itself is perennial cannot be employed on short-term contracts so as to make them feel insecure and discourage them from organising. There are such laws in some countries, for example the Contract Labour (Regulation and Abolition) Act, 1970, in India, but most trade unions have not fought sufficiently hard for this kind of protection, perhaps regarding these underprivileged workers as marginal or being unable to mount the campaign required. The result is that supposedly marginal workers have, in many countries, become the majority of the labour force.

From the standpoint of retail and brand-name companies, most irregular workers as well as their employers are likely to be invisible. First-tier employers may not admit that they are subcontracting out part of their production, since they can make an extra profit from the lower wages paid to these workers; and the sheer difficulty of tracing the supply chain through long and shifting links (sometimes mediated by agents) will be enough to put off even conscientious independent monitors. Perhaps the biggest problem we faced in our research was at this lower level, where the question 'Who is producing how much for whom?' could rarely be answered in full.

Despite all these problems, however, there are still ways in which codes of conduct can be used to help this category of workers. Retailers and brand-name companies would have to be required to check production in each of their first-tier suppliers against the number of workers employed by them to find out if subcontracting was taking place, and if so, take responsibility for monitoring the next tiers down the chain. In theory this is supposed to happen, but in practice all our country case studies found that monitoring at lower levels of the subcontracting chain was non-existent. In addition, retailers would need to ensure that, in each tier, workers who could be employed on a permanent basis were not being employed on temporary contracts in order to make them feel insecure and thus preclude efforts to build union organisation. They would have to undertake that the prices they paid suppliers were sufficient to allow the latter to pay temporary workers properly and employ permanent workers wherever possible. And they would have to establish long-term, stable relationships with their suppliers rather than chopping and changing. Suppliers are understandably reluctant to invest in improvements such as higher wages and better working conditions if they are then likely to be dumped by their buyers in favour of others who can offer their goods more cheaply, and they would certainly not be willing to recognise and negotiate with unions if they run the risk of being dumped by their buyers if and when they are unable to complete an order on time due to a strike or go-slow. Suppliers need stability in their relationships with buyers if they are to provide employment stability to their workers.

All this adds up to a massive task, and, if carried out conscientiously, would be so expensive that it would negate the cost-effectiveness of subcontracting in the first place. If retailers and brand-name companies had to take responsibility for *all* subcontracted workers, the end result would probably be a stabilising and shortening of the chain in order to cut the costs of monitoring and, indeed, this is precisely what happened in the supply chains of some of the retailers who signed up to the base code of the Ethical Trading Initiative (S Porto, personal communication, 2004). At first sight, this might be seen as a disastrous outcome for workers lower down the chain, who would lose their jobs. But if we assume that the quotas of the suppliers remain the same, they would still need to have the same number of employees working for them, only, instead of most of the workers being employed via sub-subcontracted sweatshops, most would be directly employed. As our research showed, the higher in the chain workers come, the more likely they are to be able to fight for their rights, and the more possible it is for others to help them. So the overall result would be an improvement in employment conditions and workers' rights. If a safety net were provided for the workers who initially lost their jobs until they found a new one, or they were helped to form co-operatives, they would not be any worse off, since their work was already precarious and insecure to begin with. There are attempts to address these concerns by some retailers, but the clothing industry is still driven by competitiveness which often translates into price and time pressure on suppliers, which is then passed down to the workers.

Informal Workers

These workers are the least visible: indeed, as legal entities they do not exist at all. They have no contract, no proof of employment and no recognition of their status as workers. In the garment industry they are concentrated at the lower end of the chain, but may be present at all levels, including the first tier. And the overwhelming majority are women.

The simplest strategy for employers is to employ informal workers without a contract. In China, for example, migrant workers from interior rural areas were being used in this way. Our research found that: 'They receive no legal minimum wage nor OT compensation. They work long hours, have no legal social security provisions, no contract and can be dumped at any time during the low season. Their migratory character makes their labor as 'flexible' and unprotected as the homeworkers' (Women Working Worldwide 2003a:155). This strategy is used widely, at all levels of the supply chain. However, it is relatively rare in first-tier factories, while lower down the chain, especially in sweatshops, it is standard practice. In the case of homeworkers, even those who have direct contact with the employer—those who actually go to the factory to pick up work and go back to return it when it is completed, like the homeworkers doing finishing work for Patel Hosiery in Bombay—it is almost unheard-of for them to have any proof of employment, so virtually all of them would belong in this category.

The use of a labour contractor as an intermediary between employer and worker is another way of avoiding the employer's legal obligations to workers. This is a common practice in India and Pakistan, for example. The so-called 'contract workers' (who are really 'no-contract workers') may work at the factory or sweatshop itself, yet they are not registered as employees either of the owner or of the contractor—in other words, they have no legal employer. Often homeworkers too receive work via a contractor or agent, and, needless to say, they are not registered as being employed either by the contractor or by the factory or sweatshop. Here there is an additional problem, since very often the workers have no idea who their real employer is, much less where their orders come from. Perhaps most damaging of all is the system whereby a supervisor or line leader at a factory takes work home and subcontracts it to women or families in the neighbourhood. This was found in Thailand, where supervisors became subcontractors (Women Working Worldwide 2003a:143), and in the Philippines: 'The line leader is a regular worker of the company who supervises each assembly line. The products from the assembly line need to be hand-sewn for closing tips, linking, bind-off and hemming. The line leader subcontracts these various processes of the production line to home-workers' (Women Working Worldwide 2003a:187) Here, the fact that the homeworkers are paid a miserable wage while many of the line-leaders-cum-subcontractors 'are relatively well off compared to the other regular workers during the peak season since they earn additional income from subcontracting' (Women Working Worldwide 2003a:187) causes bitterness and tensions within the workforce. The excessively low rates paid to homeworkers also results in children, especially girls, becoming workers (Women Working Worldwide 2003a:188).

The extreme oppression suffered by informal workers has become more and more widely recognised in recent years, but the problem of how this injustice can be addressed has yet to be solved. It is necessary to investigate why this is so, and what has so far been done in this area, before coming back to the proposals made by the women workers in our research.

What is informal labour?

In discussing informality, we face the problem of definition. For a long time, everyone casually referred to 'the informal sector,' and many people still do. This consists of small-scale, unregistered enterprises, which may or may not employ paid workers. An aura of romanticism hangs around the term, either because it is thought that 'small is beautiful,' or because it is felt that this type of enterprise is more suitable for developing countries, and/or provides a stepping stone from rural labour to urban formal sector employment. Although this conception has been criticised, it still prevails in some quarters, and makes the task of fighting for the rights of informal workers that much more difficult (Banerjee 1981). In the first place, it blurs class boundaries. No one would think of clubbing together corporate CEOs, professionals like doctors and lawyers, and shop-floor workers under the umbrella term 'the formal sector,' as though they had interests in common, yet precisely this is done when employers, the self-employed and workers are clubbed together under the term 'informal sector.' In the second place, it leaves out the majority of informal workers, including those who work in large-scale formal enterprises, or as homeworkers, domestic workers or rural labourers.

Part of this criticism is addressed in the shift to the term 'informal economy', popularised especially by WIEGO (Women in Informal Employment: Gobalising and Organising). According to their booklet *Addressing Informality, Reducing Poverty: A Policy Response to the Informal Economy* (2001):

> the defining characteristic of the informal economy is the precarious nature of work: workers in informal enterprises and informal jobs are generally uncovered by social security or labour legislation...All non-regular wage workers, both rural and urban, who work without minimum wages, assured work or benefits, and virtually all self-employed persons (except professionals and technicians) belong to the informal economy. (WIEGO 2001:1)

Without actually being said, it is also implied that there are employers who belong to the informal economy, and, indeed, it would be absurd to have an *economy* that includes waged workers but not employers.

The main virtue of this definition is that it includes a huge swathe of informal workers who work outside the 'informal sector' as defined solely in terms of the nature of enterprises. But it still lumps together employers, the self-employed and workers, and especially the latter two categories. In this it follows quite closely the terminology of SEWA (Self-Employed Women's Association, based in Ahmedabad, India), which also speaks of 'home-based' women workers, not distinguishing self-employed workers from homeworkers.

A drawback of this approach, however, is the confusion it causes about a worker's employment status. Today's homeworkers are in all essential respects in the same position as nineteenth-century workers in domestic industry, which Marx described as 'an external department of the factory, the manufacturing workshop or the warehouse...An example: the shirt factory of Messrs. Tillie at Londonderry, which employs 1,000 workers in the factory itself, and 9,000 outworkers spread over the country districts' (Marx 1977:591). Perhaps the most important difference is that today, however, workers have rights, including social security, which makes it even more convenient for an employer to make a worker believe she is self-employed, since he is then relieved of all the responsibilities of an employer. Our research in Sri Lanka found that 'the most important aspect we were able to learn from the surveys was that workers are being told they are self-employed by their employers, in order to avoid their legal obligations of paying EPF etc, and renege on all other employer responsibilities towards workers' (Women Working Worldwide 2003a:82). Rejecting such a conception, the homeworkers in Bombay were very clear that they were waged workers; as one of them said, 'Just because we work at home, that doesn't mean we're not workers. We *are* workers, and we should get all the same rights that factory workers get' (comment made during URG workshop, Bombay, 2002). One of the grievances they listed was that they had to pay for their own thread and thread-cutters, which they (quite rightly) regarded as a deduction from their already meagre wages. It may not always be easy to draw the line between self-employed home-based workers and homeworkers, but it should be obvious that women who stitch just one part of a garment, or do only finishing work, are not really self-employed. The crucial question is that of access to the market: if it is not the worker herself who sells the product to the customer, then the person who pays her is really her employer. And while own-account workers undoubtedly suffer serious problems, especially due to lack of access to social security, there is still a critical difference from wage workers, in that there is no employment relationship.

The WIEGO booklet rightly criticises the obsolete view that 'The informal economy is the traditional economy that will wither away and die with modern industrial growth,' pointing out instead that 'The informal economy is increasing with modern industrial growth' (Women in Informal Employment: Globalizing and Organizing 2001:2). This lays to rest the optimistic view that informal workers can automatically look forward to secure employment in the formal sector, but stops short of explaining why the traffic is in fact going the other way—from the formal to the informal sector. Research in India does not support the view that informal production is at all more efficient—on the contrary, the subdivision of workforces into smaller and smaller units makes it difficult or impossible to upgrade technologically, and results in less efficient, less competitive production (Krishnamoorthy 2004). Nor is the idea that informal enterprises are more labour-intensive than formal enterprises and therefore likely to create more employment borne out by research, which shows that often the two sectors use the same technology (see for example, Holmstrom 1984:151), and demonstrates that if it costs much less to create jobs in the informal economy than in the formal economy, this is mainly because the outlay on wages, working conditions and overheads is so much lower in the former. Indeed, as our research shows, exactly the same work can be done in a factory, a sweatshop or at home.

So why is informal labour increasing? The research from Sri Lanka came up with an interesting correlation: 'we can see a growing trend that when a factory organises [ie the workers unionise] the manufacturer wants an escape route to subcontract to smaller factories and micro-enterprises that are not organised' (Women Working Worldwide 2003b:50), suggesting very strongly that the purpose of using informal labour is to avoid dealing with a unionised workforce. Our research in Bombay confirms the view that the real purpose of informalisation is to deprive workers of their rights. In one sweatshop, where workers were being paid R850 per month (less than a third of the minimum wage) they formed a union and presented a charter of demands. Instead of negotiating, the employer closed down the unit, restarted it after two months, and took back the workers on a wage of R800 per month (Women Working Worldwide 2003a:99). The case of Prakash Garments, where informal workers tried to unionise is also typical. 151 workers joined the Sarva Shramik Sangh union at the beginning of 2002, and on 3 April the employer declared a partial lock-out affecting just these 151 workers. The union took the case to court, but had difficulty establishing that it was an illegal lock-out because the workers had no proof of employment. In some cases, this opposition to unionisation is because a unionised workforce means higher wages and less 'flexibility' for the employer, while in others it may be that the employer is ideologically opposed to workers having rights. Either way, the only real 'advantage' to the employer is that informal workers have no power.

Once it is recognised that, by definition, informal jobs are not covered by labour legislation or social protection, and, consequently, informal wage workers do not enjoy secure contracts, minimum wages, worker benefits such as paid leave, off-days or holidays, social protection for illness, disability and old age, nor, most importantly, the right to organise, it is evident that proposals for informal workers such as the proposal for social security schemes covering all informal workers made in India by the second Report of the National Commission on Labour (Ministry of Labour 2002) are not the solution. This may sound good, but fails to identify the basic problem as informality itself; unless workers have proof of their identity as workers and proof of employment, however temporary it may be, they will not be able to claim the rights and benefits that they are entitled to. The real purpose of the proposals for informal workers in this report appears to be to act as a sugar coating on a bitter pill, because it also proposes to eliminate regulation of closures and dismissals in the formal sector, as well as restrictions on the use of contract labour, from the existing labour laws (Ministry of Labour 2002). This would propel workers currently in formal employment into informal employment, and block the route available to contract workers attempting to formalise their employment, so that the overall consequence would be the spread of informality and further loss of workers' rights.

The somewhat schizophrenic approach which on the one hand acknowledges that informal workers lack rights, but on the other hand extols the virtues of the informal sector, can be traced to a confusion, once again, between the place of work (large or small, factory or home) with the type of employment (formal or informal)—the very confusion that the use of the term 'informal economy' was supposed to preclude. This confusion cannot be avoided unless it is made very clear that we are referring to informal *workers* in informal *employment relationships*. And as a conference on informal workers organised by IRENE (International Restructuring Education Network Europe) and attended by 62 women from all over the world noted, the fundamental problem of informal workers is their invisibility, and therefore, when defending their rights, 'the pivot is visibility' (Women's International Network 2002).

The *Resolution Concerning Decent Work and the Informal Economy* of the General Conference of the ILO, 2002, deplores the fact that 'Unregistered and unregulated enterprises often do not pay taxes, and benefits and entitlements to workers, thus posing unfair competition to other enterprises' (point no.12). It correctly identifies the problem of informality as 'principally a governance issue' (no.14), emphasising that 'the legal and institutional frameworks of a country are key' (no.16), and that therefore 'the government has a primary role to play' (no.21). The resolution insists: 'The elimination of child labour and bonded labour should be a priority goal' (no.22), pointing out that 'Child labour is a key component of the informal economy. It undermines strategies for employment creation and poverty reduction, as well as education and training programmes and the development prospects of countries' (no.22).[4]

Given all these evils of the informal economy, the ILO draws the logical conclusion that 'programmes addressing the informal economy... should be designed and implemented with the main objective of bringing workers or economic units in the informal economy into the mainstream, so that they are covered by the legal and institutional framework' (no.25), and adds that 'In many developing countries, rural development and agricultural policies, including supportive legal frameworks for co-operatives, need to be enhanced and strengthened. Special attention should be given to the care responsibilities of women to enable them to make the transition from informal to formal employment more easily' (no.26). In essence, this is also the conclusion coming out of our research, especially in India: the informal sector has to be eliminated, because it cannot be humanised. And the women workers made suggestions that could help the ILO in its goals to 'identify the obstacles to application of the most relevant labour standards for workers in the informal economy' (no.37c), and 'identify the legal and practical obstacles to formation of organisations of workers... in the informal economy' (no.37d).

Defending the rights of informal workers

So what are the obstacles faced by informal workers when they try to organise and fight for their rights? The ILO resolution suggests that 'Trade unions can sensitise workers in the informal economy to the importance of having collective representation through education and outreach programmes' (no.34), and this is exactly what we did in Bombay as part of our project. However, it was clear to us that lack of awareness was the least of the problems faced by workers. They were quick to learn, and some, indeed, were not only aware of the need for collective representation, but had already made one or more attempts to act on this knowledge. The problem was that they then lost their jobs, and all of them knew that if they tried again, the same thing would happen again.

Enrique Marin of the ILO neatly summed up the dilemma faced by these women workers when he said that 'Certain employers have found they don't need to have the law changed; they just side-step it by changing the employment relationship. They vacate the law' (Dutch Trade Union Federation 2003:19). In some cases labour legislation is at fault, because it defines 'employees' or 'workers' (or, more anachronistically, 'workmen,' as in India) who have the right to unionise in such a way that informal workers are not covered. In such cases, a more inclusive definition is required. Moreover, since the ILO Conventions on Freedom of Association and the Right to Organise and Bargain Collectively apply to *everyone*, from employers to workers, there is no argument for excluding anyone by claiming that she is self-employed. However, for other rights (such as minimum wages, social security, paid leave, and so on), it is important to distinguish between employees and those who are genuinely self-employed. Labour legislation needs to be formulated in such a way that it is impossible, or at least very difficult, for employers to claim that waged workers are self-employed.

The problem still remains, however, that the legal right to organise and bargain collectively means nothing if you can be dismissed for attempting to use that right. If you have proof that you are a worker and were working for a particular employer, you have some chance of legal redress if you are dismissed illegally. But if not, you have none. This is why two of the main demands made by informal workers in Bombay whom we spoke to for this research were firstly proof of their identity as workers (they suggested photo-identity cards and registration), and secondly proof of employment (they suggested attendance diaries which would be stamped by the employer or employers every time they received work). Since the problem is compounded by the fact that many of the employers themselves are not registered, or work is distributed through labour contractors so that there is no direct relationship between workers and employers, we also suggested at the end of the discussion that all employers and workers in the garment industry should be registered with a tripartite board consisting of representatives of employers, workers and government, and that all employment should take place through the board, so that it would have a record of who employs whom for how long, and ensure that workers get proper wages and benefits and are not victimised for organising. Once employment has been formalised in this way, it would be possible to apply the conditions suggested for irregular workers above.

This may sound impossibly utopian, but it is not. In India there are already laws supporting the formation of tripartite boards with the task of registering all workers and employers in other industries—for example, the Dock Workers (Regulation of Employment) Act, 1948 and Maharashtra Mathadi, Hamal and Other Manual Workers (Regulation of Employment and Welfare) Act, 1969. And in Bulgaria: 'A new law has very big penalties for employers who don't register, to push employers to employ workers formally', and if informal workers 'can use civil law to prove they have in fact been in a real labour relationship with the employer for more than 3 days', the employer has to keep the workers in their jobs and pay them back wages if they were dismissed (Women Working Worldwide 2003b:60–61). With the advent of computers, registering all workers and employers and keeping track of employment relationships is quite feasible, and this is what we must press for.[5]

Can codes of conduct help in this process? If informal labour is being used by first-tier supplier factories—as in the case of the factories in Guangdong, China, and Patel Hosiery in Bombay—it is possible to use codes which require a proper contract and employment relationship with all workers to insist that the employer formalise their informal workers. Further down the chain, with both employment and subcontracting relationships shifting so frequently that by the time it is established that a worker has been working for some retailer she is no longer doing so, codes cannot help very much. However, it is still worth insisting that such clauses should be included in codes and their implementation monitored carefully, since this would help to eliminate informal labour in first-tier units. Similarly, collective agreements signed by unions representing workers in retail companies should try to include such clauses pertaining to supplier factories, which would enable their union to intervene legally if they find that informal workers are being employed in supplier units.

At lower levels of the chain, however, the role of the government becomes crucial, so that putting pressure on employers alone cannot solve the problems. Where child labour is concerned, for example, the most that can be done through employers is to ensure that they do not employ children, and pay their employees adequate wages so that they do not have to send their own children to work. But if a retail or brand-name company discovers that large numbers of children are being employed by one of its suppliers, what can it do to ensure that these children's livelihoods and education are taken care of? This cannot be achieved without the co-operation of the government. Indeed, even the task of registering all workers and employers has to be undertaken by the government. And while monitoring by buyers and independent monitors can be effective in a limited number of units, defending workers' rights on a large scale cannot be achieved without government intervention. The deeper we plunge into the murky depths of subcontracting chains, the less sense it makes to defend the rights of workers as employees of a particular company, and the more important it becomes to defend their rights simply as garment workers.

Here the role of the ILO could be crucial. Over the last ten years, it has been taking informal labour far more seriously, and if it actually affirms in a Convention the right of all workers to legal registration as workers, and to proof of employment no matter how transient it may be, this would be a valuable tool in defending the rights of workers who currently cannot even think of organising without losing their jobs. Moreover, if it offers to monitor the implementation of a workers' rights clause in the WTO that includes this right, it would put to rest much of the anxiety that such a clause could be used against developing countries, and give teeth to a measure that would empower millions of women workers. Finally, the ILO conclusions include a reference to rural co-operatives, but why only rural? The women homeworkers in Bombay thought that having a production co-operative would be the only way to escape from the irregularity of their work, which left them without an income for long stretches. Moreover, having service co-operatives providing services such as cooking, cleaning and childcare would also be of great use to women workers, some of whom become homeworkers simply because they have to be at home to handle these tasks, while others who go out to work have an inordinately long working day because they have to perform many hours of domestic labour before and after their waged work. Legislation to facilitate the formation of both production and service co-operatives would be an important component of defending the rights of women workers, and the ILO as well as consumer groups should support efforts to secure such legislation in future.

Conclusion

Although workers at all levels in subcontracting chains suffer serious difficulties in getting their rights respected, strategies for addressing these problems vary depending on (a) their place in the chain and (b) their employment status. The more visible regular workers at the top of the chain can use collective bargaining, labour law, solidarity action from other workers along the chain and codes of conduct to defend their rights. For irregular workers and those who are further down the chain, this is not so easy; changes may have to be made in labour law in order to protect them from dismissal for union activities, and consumer campaigners may need to press for stricter monitoring further down the chain. For informal workers, including homeworkers, most of whom are at the bottom of the chain, the first requirement is proof of employment, without which they cannot secure any other rights. The role of the government is crucial at all levels, and where a government is unable or unwilling to defend the rights of workers, the ILO could play an important role in helping or putting pressure on the government to protect the fundamental rights of workers. Finally, a carefully formulated workers' rights clause in multilateral trade agreements of the WTO would ensure that trade liberalisation does not continue to lead to a worldwide lowering of labour standards in the industry as a whole.

Notes

1 For debates on Codes of Conduct as a means of defending workers' rights see, for example, CCEIA 2000
2 I hasten to add that this was *not* on our suggestion! In previous workshops we had discussed the importance of unions, in the context of the negative experiences some workers had had with unions that subsequently sold them out. We emphasised that this was not a feature of all unions, and that democratic unions controlled by the workers themselves could improve employment conditions considerably. But we also pointed out the difficulties and dangers, especially for informal workers.
3 Social Accountability International's first social accountability system, SA8000, is a way for retailers, brand companies, suppliers and other organisations to maintain just and decent working conditions throughout the supply chain. The SA8000 standard and verification system is a credible, comprehensive and efficient tool for assuring humane workplaces.
4 There have been arguments that child labour is a consequence of poverty and cannot be eliminated without first eliminating poverty. The opposite case—that elimination of child labour can be achieved through the provision of compulsory elementary education for all children, and that there is a demand for education even among the poorest of the poor—has been argued in theory and put into practice by Magsaysay Award-winner Shantha Sinha (See Sinha 1996).
5 It is possible that some workers in countries like the UK, which provide welfare benefits, may not wish their employment to be recorded in case they lose their benefits and/or have to pay taxes, but for workers in Third World countries like India, which have little or no social security, and where they would be earning well below the income tax limit, they have nothing to lose and a great deal to gain from registering.

9

The Phase-Out of the Multi-Fibre Arrangement from the Perspective of Workers

Angela Hale with Maggie Burns

Introduction

The changing structure of garment industry supply chains, reported in other chapters of this publication, has taken place within the context of another dramatic change in the garment industry, the phase-out of the Multi-Fibre Arrangement (MFA). The MFA is the international trade agreement that has regulated the industry for the past 30 years, and it has had a huge impact on the geographical location of supply chains. The phasing-out of this agreement is already beginning to change the global distribution of the industry and this will have implications for workers everywhere. In this chapter we will look at the background to these changes, whose interests they are serving and how they are related to the increase in international subcontracting. We will then focus on what these changes mean for workers, particularly in those countries predicted to experience the greatest loss in trade. Finally, we will look at labour movement responses and the prospects for developing multi-stakeholder initiatives to address the potential loss of livelihood amongst millions of workers.

The MFA Phase-Out

World trade is increasingly conditioned by the rules of the World Trade Organisation (WTO). In the case of the garment industry this means the phasing-out of the Multi-Fibre Arrangement (MFA), which has dominated trade in textiles and garments since 1974. Through the MFA the United States, Europe and Canada imposed import quotas to protect their domestic industries. Special measures were seen as necessary because the labour-intensive nature of the industries meant that it was becoming relatively easy for developing countries to compete. Strict quotas were established on imports from strong producers such as Hong Kong, whereas fewer restrictions were applied to poorer countries. Such selective protectionist policies cannot be sustained within the framework of the free-trade agenda promoted by the WTO. This agenda was clearly set at the Uruguay Round of the General Agreement on Trade and Tariffs (GATT) in 1986, which established the basis for the WTO. The removal of the MFA was seen by many developing countries as one of the few gains of this new trade agenda and was one of the reasons for signing up to other GATT agreements. The framework for the phase-out was provided by the Agreement on Textiles and Clothing (ATC), which would remove all quotas by the end of 2005 (see Box 9.1).

In reality the MFA phase-out is being managed by the USA and EU. Agreement was gained on a ten-year transition period between 1995 and 2005, which was intended to allow time for industry adjustment. However, industrialised countries have been able to choose which products they integrate at the various stages of the process. They have used this to deliberately hold back on changes in the items that would make a significant difference to developing countries. For example, they added items to the initial MFA list and then liberalised items not strategically important such as dolls clothes, parachutes and seat belts (Baughman et al 1997). The US published product details for the whole period, which made it clear that the intention was to 'end-load' the programme so that most items of significance are only included in the final stage. At the same time, the remaining protective measures built into the agreement have been used by the US and EU to hold up the phase-out. This includes transitional safeguards, anti-dumping measures, and rules of origin (see Box 9.1).

In the light of these developments it is not surprising that criticism of the ATC has focused, not on the agreement itself, but on the manner in which Europe and North America have controlled and circumvented its implementation. The demand has been for a properly ordered and fair phasing-out of the quota system. The current WTO Director-General, Dr Supachai, has himself called for change. He stated that only 20% of the products under the ATC subjected to quotas had so far been integrated into WTO rules leaving 80% of quotas to be eliminated. He recognised the potentially disruptive implications of

Box 9.1 The Agreement on Textiles and Clothing (ATC)

A new multinational agreement on trade in textiles and clothing was introduced as part of the GATT Uruguay Round in 1986. The aim was to bring rules on textiles and clothing trade into line with the free trade policies of the newly established Word Trade Organisation. The ATC provides a framework for the phasing-out of the Multi-Fibre Agreement (MFA) by the end of 2005.

Since 1974 world trade in textiles and garments has been governed by the MFA. This provided the basis on which industrialised countries have been able to restrict imports from developing countries. Every year countries agree quotas that detail the quantities of specified items that can be traded between them. The exporting country then allocates licenses to firms to export a certain proportion of each quota. The ATC is an agreement to phase out the MFA over a period of ten years, beginning in 1995. It applies to all WTO members whether or not they were signatories to the MFA.

Timetable for the phase-out

The main provision of the agreement is a timetable leading to complete integration at the end of a ten year transition period, i.e. by 2005.

The agreement provides a list of all the products that need to be integrated. There are two aspects to this process:

1. Progressive integration of products
   Integration is required to take place in four stages. Each stage must include products for each of four groups:
   a. Tops and yarns
   b. Fabrics
   c. Made-up textile products
   d. Clothing.
   The choice of items within these groups is left to importing countries.

2. Progressive raising of quotas
   At each of the first three stages of integration there has to be an annual increase in the quota level of those products still under the MFA.

Box 9.1 (*Continued*)

Other provisions in the ATC

Several articles in the Agreement have affected implementation of the phase-out. They were the result of strong lobbying by the USA and EU and include:

1. Transitional Safeguard Measures (Article 6)
   Article 6 continues to permit the application of MFA-type safeguard actions. The provisions are intended to prevent a sudden rise in imports of specific products causing serious damage to the importer's domestic industry. Unlike general WTO safeguard measures, these measures can be directed at specific countries. In order for them to be introduced there has to be evidence of significant damage.
2. Reciprocal market access (Article 7)
   Article 7 links the integration process to increased access to developing countries' markets. The EU and US want greater access to overseas markets for textiles and for upmarket clothing. This was seen as the price developing countries had to pay for the phase-out of the MFA.
3. False origins (rules of origin)
   The ATC requires developing countries to demonstrate that they have adopted effective measures to prevent transhipments or false declarations of origin. Importing countries can impose sanctions (e.g. reduce quotas) in the event of evidence of malpractice.

---

this situation and has called for a pre-deadline easing of quotas (Appelbaum 2003:10). Informal consultations were held at the WTO in August 2004 on the request of Mauritius whose representatives called for an emergency WTO meeting to examine the adjustment costs from the ending of quotas. However whilst 'expressing understanding for adjustment challenges,' Dr Supachai's position is 'overall ATC implementation would bring considerable welfare and efficiency gains for the global economy as well as benefits for the consumers' (Appelbaum 2003:10). Discussions are still taking place within the WTO, but so far do not appear to be addressing the problems faced by the most vulnerable garment producing nations.

In whose interests?

Whilst the focus of debate at international forums has been on implementation, other more fundamental questions about the implications of the ATC have been relatively neglected. Even if there was full and fair implementation, the question remains as to who will really benefit from the MFA phase-out. Negotiations were based on the assumption that developing countries would gain from increased access to Northern markets and the main losers would be Northern producers. Yet the consensus now is that the negative impact will be felt more in the South than in the industrialised countries of the North. The result will not be a major increase in overall access to Northern markets, but rather a relocation of garment production between countries of the South. The MFA quota system, although operating against the South as a whole, did have the effect of guaranteeing a Northern market to a large number of poor countries. With a more open system, producers will have to compete in a global market and the result is likely to be greater concentration of production in a few low-cost locations.

As far as business interests are concerned, although the MFA was set up to protect major interests in Europe and North America, it is not in fact they who will now lose from the phase-out. This is because the garment industry has been massively restructured since the MFA was first drawn up. In the early 1970s, clothing was generally produced by relatively small manufactures and sold to local and national retailers. Since then, company mergers and increased market concentration, together with the internationalisation of production through outward processing, mean that businesses operate at a global level with little regard for national boundaries.

The MFA was in itself one of the driving forces behind this restructuring. International sourcing was a way for companies in Europe and North America to overcome the high costs of domestic manufacturing in the context of a protectionist regime. Meanwhile, the strongest producers in the South, such as Hong Kong and Korea, circumvented MFA restrictions by subcontracting production to countries with less stringent quota restrictions. In both cases, only larger companies could manage these processes, so the MFA has not only increased the global complexity of subcontracting but also contributed to the increase in market concentration. A high percentage of quotas have, in fact, been filled by the reimportation of sewn goods by large Northern based companies. This means that, although part of the production takes place in developing countries, the process is still owned and managed by companies in EU and North America and it is they who reap most of the profits.

The garment industry is now controlled by major Northern-based retailers and brand based companies, together with large first tier manufacturers. These global players, whether buyers or manufacturers, can now reap the profits from the industry without any need of government protection. In fact they do not want complicated quota restrictions since these may prevent them from sourcing from the most profitable locations. Such restrictions also create extra costs and reduce the flexibility that has been made possible through computer technology and improved transport capacity. As the industry changed, it has become more advantageous to powerful business interests to remove barriers to international sourcing than to protect local manufacturing. In fact, by 1996 the Director of the German Clothing Federation announced that the motto of the German clothing industry was 'no customs, no quotas' (Textile Outlook International 1996).

Given the changing structure of the industry, it is not difficult to see why industrialised nations became willing to agree to the dismantling of the MFA. However, such positions were not stated so clearly at the beginning of the GATT Uruguay Round in 1986 when the ATC was first being formulated. A pro-protectionist position was still maintained by the EU and USA during the negotiations. Every impression was given that dismantling the MFA involved major concessions on the part of industrialised nations. This partly reflected the position of the textile industry, which still had major production facilities in the North and was lobbying hard for the MFA to be maintained. However, it was also a matter of political expediency. The MFA still had widespread support in key domestic constituencies, particularly in the US where there is a strong textile union. In the case of the EU there was an active internal lobby on the part of Portugal and Spain, which still had substantial clothing industries. Also by starting with this protectionist position the industrialised countries were better able to dictate the details of the phase-out and maintain their control over the liberalisation process.

The MFA phase-out was also used by the EU and US as a bargaining tool in the GATT negotiations as a whole. The offer to remove the MFA was an important incentive to encourage countries in the South to support other GATT agreements around intellectual property, investment and services. It was also used as a bargaining tool to persuade countries such as India and Pakistan to lower their tariff barriers. In fact increased access to developing countries' markets is built into the ATC agreement itself through the 'reciprocal market access' clause.

According to Underhill, the MFA phase-out was therefore a process through which powerful companies and their Governments were able 'to get something for nothing' (Underhill 1998:242). They no longer needed the MFA but they used the offer of removing it as a way of persuading countries to agree to the wider liberalisation agenda.

The MFA phase-out may suit major business interests in Europe and North America, but this does not mean that it is in the interests of *all* Northern manufacturers. Large manufacturers are able to compete in a globalised market through balancing increased international outsourcing with some continued investment in downsized domestic bases. However, smaller firms lack the capital resources and market knowledge to adapt to the changing dynamics of the industry. Many traditional clothing manufacturers are therefore going out of business, and those that are surviving are often doing so through subcontracting to smaller production units and homeworkers (as outlined in the case of the UK, see Chapter 6, this volume). Meanwhile the textile industry is shrinking, with 50% of US textile mills closing in the past six years (internal meeting with Trade Attaché at the US Embassy in Sri Lanka, June 2003).

Business as usual

Whilst powerful clothing companies are likely to gain from the removal of the quota system, business interests are not necessarily best served by a completely open market. Furthermore, whilst government trade representatives work closely with these companies, they also have an obligation to protect their national interest in relation to domestic production. Their agenda is to use trade negotiations to reduce the barriers to business operations in other countries, whilst at the same time maintaining as much control as possible on the entry of competitive goods into their own countries.

Since the MFA phase-out only relates to quotas, other restrictive mechanisms, such as tariff barriers and complicated rules of origin (see Box 9.1), have been increasingly used by the EU and US to maintain their control of the industry. Research has shown that rules of origin, which define the producing country as the country that adds the most value to the product, are being used to restrict the market access of some of the poorest countries (Oxfam International 2004). Meanwhile tariff barriers on textile and clothing products are three times higher than on other manufactured goods and any offers of reduction are tied to unfair demands for increased liberalisation on the part of developing countries.

In addition, the US and EU trade agenda is being implemented not only through the WTO, but also in regional and bilateral trade agreements. In fact, after the collapse of the WTO meeting in Cancun, the US openly declared that it would continue its free trade agenda through the negotiation of other agreements: 'We have free trade agreements with six countries right now. And we're negotiating free trade agreements with 14 more' (Robert Zoellick, US Trade Representative, quoted in Foo and Bas 2003). Regional trade agreements include the North America American Free Trade Agreement (NAFTA), which has tied Mexico into the US economy, the US/Central American Free Trade Agreement (CAFTA) and the Caribbean Basin Initiative (CBI). Similar agreements exist between the EU and Eastern Europe and North Africa. In the post-2005 quota-free era, it is these bilateral and regional agreements that are going to be crucial to the geographical location of the industry.

Bilateral and regional trade agreements are being used by the US and EU to bypass the WTO and establish their own rules which institutionalise their control of the industry. Although many of these agreements give preferential market access for poorer countries, even this is being used to ensure protection of US and EU business interests, notably the still profitable textile industries. For example, the US Africa Growth and Opportunity Act (AGOA), which came into effect in 2000 and has recently been extended for three years until 2007, permits many sub-Saharan countries unlimited duty-free and quota-free access to the US market for clothing made with US fabric and thread. Similarly, the Outward Processing Trade rules of the EU allow the tariff-free importing of garments assembled in East Europe and North Africa, which have used EU-manufactured fabric (Smith 2003).

Changing Global Distribution

The question dominating discussions of the MFA phase-out is: 'Which countries will win and which will lose?' (Anson 2003; Appelbaum 2003; Mehta 1997). The increasing importance of changing bilateral and regional trade agreements is one reason why this is not an easy question to answer. Also, it is not known how national governments, or companies themselves, will respond to new opportunities. Nevertheless, it is clear that more open markets will enable Northern-based companies to have greater freedom as to where they source and that comparative labour costs are a significant variable in location decisions. Other significant factors include the skill level and flexibility of the workforce, the nature of the country's financial system, sources of capital, flexible production processes, attitudes to technology and transport and communications infrastructure. Proximity to the market is significant, but does not necessarily imply that quick deliveries and cheap materials are more important than transit costs (Appelbaum 2003; Mehta 1997).

In spite of the difficulties of prediction, there is general agreement among analysts about the main countries that will gain and lose by the post-2005 regime. Those likely to gain include countries that lost out under the restrictive MFA regime. Overall, Asia is expected to benefit, with something like a 6% increase in trade. China is expected to gain most, already increasing its market share as quotas are removed. The WTO report that between 1995 and 2002, China's share of the US market for clothing rose from 16% to a massive 50% (Nordas 2004). Some analysts predict that by the end of the phase-out China will achieve a 150% increase in its overall textile and clothing exports (Appelbaum 2003:35), and have nearly 50% of the world market (Ianchovichina and Martin 2001). India has similarly been held back by quota restrictions and is expected to gain from the phase-out, as is Pakistan (Majmudar 1996; Nordas 2004). All have well-established industries with access to domestically produced fabrics together with low labour costs (see Table 9.1).

Table 9.1 Hourly labour costs including social and fringe benefits (US$), 1996

Country	(US$)
Japan	16.29
Taiwan	5.10
Hong Kong	4.51
South Korea	4.18
Malaysia	2.52
Mexico	1.08
Thailand	1.06
Philippines	0.62
Sri Lanka	0.41
Indonesia	0.34
Vietnam	0.32
Bangladesh	0.31
China	0.28
Pakistan	0.26

Source: Kelegama & Epaarachchi 2001

The story is different for those countries whose garments industries expanded rapidly under the quota system, notably Bangladesh and Sri Lanka. These countries had no natural resource base or historical tradition of garment exports. They mainly owe their industry to relocation by companies from other Asian countries which, having utilised their own quotas, set up factories in South and South East Asia in order to take up those countries unused quotas. Low wages and ease of entry into these countries to set up factories are other reasons for the growth. This growth has been most dramatic in Bangladesh, where garments accounted for 4% of the total national exports in 1983 and as much as 78% in 2001 (Appelbaum 2003:40). Bangladesh has poor infrastructure and, although labour costs are low, productivity is not very high. Therefore, all predictions indicate that, without preferential treatment, Bangladesh will be unable to compete against stronger players such as China and India in the post-MFA world.

The situation for many other countries is less certain. For example, in Asia the elimination of quotas is beginning to cause a decline in exports from the Philippines and Thailand, but both maintain certain advantages such as good infrastructure and a skilled workforce. Meanwhile the prospects for countries close to the US, notably Mexico, and the countries of Central America and the Caribbean, are also unclear. Production costs are higher than in Asia, and they are vulnerable to the increase in US imports from China. However, proximity to the North American market and preferential trade agreements mean that some lower-cost locations, such as Haiti and Nicaragua, will continue to be important suppliers of mass-produced garments requiring quick turnaround. Lower-wage economies in Eastern Europe will similarly benefit from proximity to the EU market and associated preferential trade agreements.

Implications for Workers

The question of who gains from the MFA phase-out can be asked not only about the distribution of benefits between different countries but also between capital and labour. However, this question has barely been raised. Most of the dialogue has taken place in the context of intergovernmental forums where the assumption is that what benefits a country benefits its population. Also, since the framework is set in terms of competing comparative advantage between regions and countries, there have been no openings for debating the global impact on the people who work in the industry. Yet this impact will be massive. David Birnbaum, an industry analyst, recently announced that millions of workers will lose their jobs (internal meeting). Meanwhile, although employment is increasing in countries which are gaining from the phase-out, the prospects for workers' rights is not good, since the comparative advantage of these countries includes low wage costs and an unorganised workforce.

Loss of employment

The relocation of the garment industry associated with the MFA phase-out clearly has implications for the global distribution of employment. In the first place there will be an increase overall in the movement of jobs from the industrialised countries of the North to developing countries of the South. It is Northern workers, not major business interests, who will experience the negative effects of the eventual phase-out. With a more open market, workers in Europe and North America will increasingly compete for their jobs with workers in low wage economies. However, this loss of jobs has been well underway for more than a decade and is more the result of global restructuring than the MFA phase-out. In 1999 The UK National Union of Knitwear, Footwear and Apparel Trades (KFAT) reported that an increase in international outsourcing was causing a loss of 1800 jobs a month from UK manufacturers (KFAT 1999). In a short period of time the union lost half its membership. Similarly, in the US, jobs in the garment industry have shrunk by just over half in the period 1990-2003, associated not so much with the MFA as with the North America Free Trade Agreement (NAFTA) which enabled US retailers and manufacturers to shift production from the US to Mexico (Foo and Bas 2003).

In fact, the MFA phase-out will cause far more job losses in countries in the South than in the North. Here the future for millions of garment workers is extremely uncertain. If current disadvantages are not overcome, there will be massive redundancies with the only question being how many and where? One thing that is clear is that Bangladesh faces a crisis, with at least one million garment workers set to lose their jobs (Lips et al 2003). It is also estimated that a further million jobs will be lost in Indonesia and 200,000 in Sri Lanka (Kearney 2003b), many in rural areas. These figures need to be considered not just numerically but also as a percentage of the total workforce. The garment industry is the biggest industrial employer in Sri Lanka and Bangladesh, accounting for

33% and 40% of the workforce respectively. As many as half of these workers are predicted to lose their jobs in Bangladesh and up to one third in Sri Lanka (see Boxes 9.2 and 9.3). Similarly losses are beginning to take place in some countries in Mexico and Central America, with 325 of Mexico's 1122 garment *maquilas* closing down since January 2001 as the US switched to imports from China (Foo and Bas 2003; see also Chapter 7, this volume).

Box 9.2   Bangladesh

The garment sector accounts for about one-third of the industrial workforce and contributes about a quarter of the gross value-added in the manufacturing sector.

Export earnings from this sector account for more than 75% of Bangladesh's total export earnings. As a result, the foreign exchange reserve of the country largely depends on the ready-made-garment sector.

At present under the MFA Bangladesh has a guaranteed market but once the quotas are removed it will find it hard to compete.

The impact on the industry will be huge and result in a great number of factory closures and much unemployment. The UNDP claims that up to a million jobs will be lost.

However, Bangladesh does have some preferential treatment because it falls within the definition of a least developed country (LDC).

Box 9.3   Sri Lanka

In 2003 the garment industry contributed to 50% of the country's export revenues and accounted for nearly 5.5% of the county's GDP.

The industry is the largest industrial employer in Sri Lanka, with a strong bias toward unmarried female workers.

Although predictions vary, it is predicted that the MFA phase-out will result in a serious loss of employment and revenue to the country.

Losses are expected to be greater for small and medium-sized firms than for bigger companies.

One of the main strategies of the Government is to try to achieve an advantageous free-trade agreement with the US.

The human cost of the MFA phase-out is therefore enormous. The vast majority of the garment workers set to lose their jobs are women who migrated from poor rural areas to seek a source of livelihood not only for themselves, but for their family as a whole. They work long hours in unacceptable conditions, but can at least support themselves and have gained some measure of personal freedom (Kabeer 2000; Kibria 1995). If the industry collapses, what will happen to these women? It is questionable whether they will be able to reintegrate back into their villages. Their lives in the factories go beyond the bounds of what is traditionally acceptable for a future wife. In Sri Lanka, for example, marriage adverts announce 'factory girls need not apply.' In Bangladesh, where garment workers have transgressed the boundaries of Islamic tradition, the stigma is even greater. But if they stay in the city, how will they survive? And what will the impact of their loss of earnings be on the poor rural areas they come from? Similarly, for working-class urban families, women's income from garment work has been vital for family and community survival, so it is not just workers themselves who will suffer.

Lowering of standards

Workers in countries which are expected to benefit from the phase-out, such as China and India, will see an expansion of employment opportunities. However it is important to consider what this implies in terms of labour standards. One the main reasons why the industry is expected to relocate to these particular countries is the low cost of labour. China is predicted to be a major supplier of the world market largely because it is seen as having a seemingly endless supply of cheap labour. For workers this means not only minimal wages, but also a highly disciplined and intensive work schedule. In garment factories supplying the world market it is normal for notices to be posted up listing fines and physical punishments for such offences as speaking or drinking water during working hours, sitting or resting and going to the toilet too often (China Labour Education and Information Centre 1996). Women garment workers typically work 10-14 hours a day, seven days a week and gross neglect of industrial safety has resulted in numerous factory fires. Furthermore, workers are prevented from organising in defence of their rights by the lack of freedom of association.

In the case of India, the other country predicted to gain, the garment industry is sustained by thousands of women workers in small scattered workplaces. Evidence presented elsewhere in this book demonstrates that it is amongst these workers that conditions are worst. In fact it has been argued that the growth of this large informal sector in India is due partly to employers preference for unregulated labour (see Chapter 8, this volume). The MFA phase-out is predicted to be accompanied by a trend away from these units and a move towards larger units in Tamil Nadu and Delhi. In so far as labour rights are concerned, this could be viewed as a positive trend. However, even in these bigger factories most workers do not have the status of permanent employees and can be dismissed without notice, making it very difficult for them to claim even basic rights. This shifting of employment within India also demonstrates that even in countries which are likely to gain, garment workers cannot be sure that their own jobs will survive.

The phase-out of the MFA is likely to be accompanied by a downward pressure on labour conditions everywhere, including in industrialised countries of the North. A more open market means increased competition and therefore pressure on both countries and manufacturers to reduce costs. In a private discussion in June 2004, a supplier in Sri Lanka reported that a buyer with a large UK retailer stated that if he was buying a garment for US$10 he expects to get the same garment at US$5 the following year (2005). In a labour-intensive industry like garments, this means labour costs have to fall. As we have seen, one way of cutting costs is to subcontract to small work units and homeworkers where labour protection is weakest. This is happening not only in parts of the South but also in the EU and US where there is also increased pressure to accept new flexible work patterns such as irregular shifts without adequate overtime pay (Christerson and Appelbaum 1995; see also Chapter 6, this volume). Marceline White, who works with Asian and Mexican immigrant women in the US, believes that the MFA phase-out is accelerating the deterioration of working conditions that garment workers have already been experiencing over the past decade. Workers complain of low pay and extremely long hours. Some have been active in campaigns against overwork, where they were forced to work in excess of 100 hours a week without overtime pay, with increased physical injury and stress (internal communication from White, 1999).

In Europe the more open trading system meant that by 1996, Poland had already become the sixth-largest clothing supplier to the EU. However, in Poland wages are kept so low that women have to work long hours to reach excessively high targets in order to earn wages that are below the legal minimum (Clean Clothes Campaign 1999:10). Poland's entry into the EU trading bloc, which occurred on May 1 2004 is intended to boost Poland's economic growth, but some warn that it may drive up costs as the world's garment industry braces itself for intensified competition associated with the phase-out (Women's Wear Daily 2004). In Eastern Europe, as in Asia, production shifts partly in accordance with differences in wage rates. Sourcing has moved from Poland and Hungary to Romania, Slovenia and Bulgaria (Begg et al 2003). In Bulgaria workers have reported that they are constantly threatened with the need to keep down production costs in order to compete. This competitive threat is often the justification given for excessive unpaid overtime and for locking workers in factories until targets are completed (Clean Clothes Campaign 1999:23; see also Musiolek 2004).

Labour Movement Responses

Trade unions and labour NGOs have found it difficult to respond to the situation that will face workers at the end of the MFA phase-out, and have been somewhat slow to react to what many now see as an impending crisis. One of the problems is that the complexity of the issues and the remaining uncertainties about outcomes make it difficult to even understand what is at stake, let alone deal with it. Furthermore, mainstream concern at both national and international level has centred around different countries' 'national interest,' rather than the issues facing workers. Consequently, the priority for labour activists in countries such as Bangladesh has been to work together with industry and the government to 'save the industry,' though trade unions are also drawing up recommendations specifically in relation to workers' rights. For Northern-based NGOs, such as Oxfam, the focus has been on the implications of the maintenance of Northern protectionism (Oxfam International 2004), though Oxfam International is also working with labour rights organisations in vulnerable countries, notably Bangladesh and Sri Lanka. There are, however, important issues which cut across all countries and which relate to the increased power of textile and clothing multinationals to exploit the labour of vulnerable workers everywhere. The situation facing workers needs to be separated out from questions relating to the viability of particular national economies.

It is the International Textile Garment and Leather Workers Federation (ITGLWF), the global federation of trade unions in this sector, that has now taken up the issues most clearly from the point of view of workers in the industry as a whole. It also works in liaison with some labour NGOs, such as the Maquila Solidarity Network, which take a similar perspective. Also as WWW we have also produced articles and activists' briefing papers highlighting the way in which the MFA phase-out is increasing the overall downward pressure on labour conditions (Hale 2000b, 2002). Since the MFA phase-out is taking place within the framework of the WTO, a central activity of the ITGLWF has been lobbying at the WTO and associated trade meetings. In the lead-up to the Cancun round of WTO meetings, the ITGLWF proposed a multifaceted strategy to address the fallout from the MFA phase-out, combining labour rights, trade and industry upgrading proposals. This included the continuation of some targeted trade restraints beyond 2005, the inclusion of labour standards conditionalities in bilateral and multi-lateral trade agreements and support for emerging and struggling industries. It also recommended national industrial policies such as industry upgrading, skills training for worker, and promotion of respect for international labour standards. Furthermore, the international trade union movement called upon the meeting to instruct the WTO Trade and Development Committee, together with the UNDP, UNCTAD, ILO and World Bank to undertake an urgent review of the impact of the MFA phase-out on development sustainability, employment and working conditions in the textile and garment sector (Kearney 2003b). Given the collapse of the talks, this proposal was not addressed.

Since the collapse of the WTO meeting in Cancun, some labour movement activists have diverted attention to the inclusion of labour standards conditionalities in bilateral and regional trade negotiations. The aim is to make labour rights compliance an essential element of a country's or region's 'competitive advantage' (Maquila Solidarity Network 2003c). In fact, both the US and EU already have mechanisms for lowering tariffs on the basis of adherence to international labour standards. For example, the US provides Cambodia with preferential treatment on the basis of trade union rights and participation in ILO inspection programmes. In practice however, labour rights clauses in trade agreements are often cosmetic. For example when the US/Central American Free Trade Agreement (CAFTA) was being negotiated recently, some labour movement NGOs lobbied actively for alternative proposals that would address Southern concerns about both market access and labour rights issues. A labour component has been included in the final text, but according to Human Rights Watch: 'It fails to require compliance with even the most basic internationally recognised labour norms and . . . whilst it calls on countries to uphold their own labour laws . . . it provides a weak enforcement mechanism for that limited commitment. . . .' (Human Rights Watch 2004). There are also those who criticise the linking of trade and labour standards because it is seen as operating against some of the poorest producers (see also Chapter 8, this volume).

Whilst labour movement activists are busy lobbying at international forums, workers themselves remain unaware of the impending changes. Millions of garment workers continue to sit at sewing machines completely unaware that a major catastrophe is about to happen. Most are used to working in a situation of day-to-day insecurity, but in countries like Bangladesh and Sri Lanka, which have geared their economies around garment production, they have also become used to the idea of being able to leave one factory and find a job in another. No-one has told them that within the next few years many of these factories will close and the jobs disappear altogether, let alone involve them in discussions over what to do about this. It is very difficult for labour activists to deal with this situation. No-one really knows which factories will go, which jobs are vulnerable, and even if this information was available, it is not clear what purpose would be served by informing workers. When the issue was discussed at an Oxfam International workshop in 2002 one worker remarked: 'Does that mean my job is vulnerable? I always thought I work for a good factory and my job is safe.' There seems little point in undermining this relative sense of security unless there are ways of involving workers in positive proposals for constructive action or alternative employment.

A Multi-Stakeholder Approach?

The impact of the MFA phase-out on workers is difficult for the labour movement to address, not only because of the scale and complexity of the issues but also because it is beyond the scope of industrial relations. Effective action demands the involvement of other players apart from workers and employers. However, it is not clear which institutions hold responsibility. Is it the WTO? Is it the major buyers? Or is it national governments? Should the situation be addressed by the World Bank and the international donor community? In a sense, all these institutions have different roles to play but, as with the labour movement, there is little they can achieve in isolation. There needs to be a concerted and co-ordinated effort by all parties to recognise and address the coming crisis, in other words the situation needs to be addressed though a multi-stakeholder approach. The development of the multi-stakeholder principle in other forums on labour rights in the garment industry has in fact set the scene for the emergence of just such initiatives on the MFA phase-out at both at the national and the international level.

National level

In Sri Lanka a process is underway that fully involves garment worker representatives. This was initiated by Oxfam International (OI) in response to local demand. Trade unions and NGOs in Sri Lanka work well together and have formed a united forum to campaign for the rights of garment workers both nationally and internationally. However, it has not been easy to persuade industry representatives to meet with them. When OI began work on the MFA phase-out it initially met separately with industry and government representatives. It was not until July 2004 that the ILO brought all parties to the table: Government ministers, leaders of the industry, NGOs, and trade unions. A multi-stakeholder task force was set up which is focused on the implications for workers. Two issues are being addressed: how to minimise jobs losses, and how to manage retrenchment. On the issue of retrenchment one proposal is that workers be given a certificate when they leave employment so that they can be given first preference when new jobs are created. The Government has also agreed to map out the situation in order to assess where the problems are likely to be. Of course, given the different parties around the table, these negotiations are not without tensions; but in a country with a history of antagonist industrial relations it is remarkable how much has already been achieved through this multi-stakeholder approach.

International level

The problem with national initiatives is that, within the competitive environment of the MFA phase-out, gains within one country are likely to be offset by losses in another. What is needed is an initiative that takes a broader approach and looks at the impact on workers globally. Ironically, it is the big multinationals that are best placed to understand what this is likely to be, and some have begun to recognise that they have a responsibility. It was at a Nike stakeholder forum in February this year, which for the first time included the trade unions, that the idea of an international multi-stakeholder approach emerged. Nike had recently pulled out of a factory in Indonesia, which involved 7000 workers losing their jobs, and the consequent turmoil had alerted Nike staff to the possible consequences of future job losses. The implications of the phase-out for workers was hotly debated at the meeting, and by the end there was a proposal to establish a global multi-stakeholder initiative involving companies, unions, NGOs and the World Bank.

Following the Nike meeting negotiations took place to bring more players on board. All parties lobbied for others to join, partly through existing multi-stakeholder initiatives such as the Fair Labor Association and Social Accountability International and a follow-up meeting was held at the Ethical Trading Initiative offices in London. At this meeting it was recognised that the situation was viewed differently by the various parties and that the motivations for involvement were not the same. The World Bank, for example, was motivated by the recognition that if there were going to be major job losses they would be called upon to provide, so they were looking for strategies to pre-empt this situation. It was also apparent that it was not even clear what the impacts of the phase-out on workers would be. It was therefore agreed that it was necessary to carry out some research to facilitate discussion.

At the time of writing this group now involves Accountability, the AFL-CIO Solidarity Centre, ASDA-George, Business for Social Responsibility, the Co-operative Group, the Department for International Development (UK), Development Alternatives, the Ethical Trading Initiative, the Fair Labor Association, Gap Inc, the Global Alliance for Workers and Communities, the Interfaith Centre on Corporate Responsibility, the International Textile, Garment and Leather Workers Federation, Marks and Spencer, the Maquila Soldarity Network, Nike, Oxfam International, Social Accountability International, the UNDP Asia Trade Initiative, the UN Global Compact and the World Bank Group, and has adopted the title of the MFA Forum. Three pieces of research are currently being carried out to facilitate its work. The first, prepared by Accountability, is now completed, consisting of a mapping exercise of existing literature, including policy recommendations as to what both the public and private sector can do to diminish the negative effects of ending the MFA for workers and their communities. Part of the purpose of the exercise was to identify key areas of agreement and disagreement (see Table 9.2 and Table 9.3). A second piece of research involves talking to buyers and first tier manufacturers about their likely sourcing policies post MFA. The third area of research is being developed by the World Bank and involves looking at other situations where labour adjustment has taken place on a large scale and seeing what can be learnt in terms of managing the transition.

It is as yet unclear how this multi-stakeholder group will move forward from this research stage, but the aim is to formulate concrete proposals that all key players will sign up to and then to collaborate on implementation. There is also a recognised need to involve other key players who, at the time of writing, are not around the table. This includes not only national governments but also other powerful players in garment industry supply chains. Corporate involvement is at the level of the retailers/brand names, who are the lead companies but who have different degrees of power over their first-tier suppliers (see Chapters 2 and 5, this volume). Many of these top suppliers, such as Li and Fung, Mast Industries and the Crystal Group, are MNCs in their own right (see Chapters 2 and 5, this volume). The degree of power that a retailer exercises over these suppliers is strongly related to the extent to which the supplier relies on their orders. If they take more than 50% of production, they are likely to have some leverage over the location of that supplier and other manufacturers further down the chain. If not, then it is difficult for a lead company to have much control over where these companies operate. This lack of leverage is even greater in the case of companies who source through agents with no manufacturing capacity of their own, since many of these agents are unwilling to even declare the details of their sourcing strategies.

The MFA phase-out is likely to accelerate the current trend towards consolidation of the supply chain at the level of first-tier suppliers, since the associated increase in competition is likely to eliminate smaller players. Whilst this gives these manufacturing companies more power, it is a trend encouraged by many lead companies, since it means that they can reduce their numbers of suppliers and build up more balanced and sustainable relationships. Through this, they can not only increase their leverage over the production process, but also over the corporate responsibility measures to which they have become committed. This commitment is now beginning to include a socially responsible approach to the phase-out of the quota system.

It is significant that discussions amongst members of the Forum relate not only to responsible management of the phase-out itself, but also to the establishment of a framework for promoting good labour conditions in the future development of the industry. There is talk of treating the phase-out not as a crisis but as an opportunity, a catalyst for re-establishing the industry on a more rational and ethical basis. This would begin by companies collectively agreeing to incorporate ethical issues into discussions of possible relocation. If it becomes impossible to continuing sourcing from a particular country or location, then it would

Table 9.2 Summary of recommendations for action in response to the end of the MFA (public sector)

Global	Garment exporting countries
What can developed country governments and international institutions do? **Possible actions:** - Trade restraints beyond 2005. - Mechanisms to link trade with ILO standards. - Easing rules of origin for LDC exports (and exports from a broader group of vulnerable garment-producing countries?). - Technical and financial assistance to vulnerable garment-producing countries and in particular to SMEs. **Agreements:** The end of the MFA is a given. However tariffs, non-tariff barriers and bilateral trade agreements remain significant in the garment trade. **Disagreements:** *There is a conflict between those calling for reduced tariffs and those who say that reducing tariffs does not directly contribute to improving working conditions. There is also a conflict between those calling for ILO standards and trade to be linked and those who believe this is protectionist and constrains the competitiveness of developing country garment industries.*	**What can developing country governments do?** **Possible actions:** - Push for regional free trade agreements (FTAs) with the US or EU and for easing of rules of origin laws in existing agreements. - Enhance competitiveness through regional integration—both in terms of markets for products and supplies of fabric. - Ease restrictions on fabric imports and remove unnecessary red tape. - Adopt industrial policies focusing on upgrading and skills development. - Improve the general business climate—improve infrastructure, customs procedures, and access to credit, reduce corruption and bureaucratic inefficiency. - Enforce labour law and respect for workers rights. - Provide social safety nets and ensure the payment of legally required social security/ pension payments and redundancy pay. Suggestions include requiring a bond to cover these payments for foreign investors. - Provide retraining programmes and job banks for retrenched workers - Ensure national legislation allows for workers to be paid before other creditors in case of closure. **Agreement:** *All commentators agree that governments in vulnerable countries should concentrate on productivity and quality enhancement and on developing workforce and management skills. A number note that countries with good labour law and enforcement could effectively market themselves to the more socially conscious US and EU brands and retailers* **Disagreement:** *There are different perceptions of the effectiveness of FTAs with some countries seeing them as the solution to their problems and some commentators saying they are overrated. Workforce flexibility and labour conditions are often seen as in conflict, but are not necessarily at odds. World Bank, ILO and industry commentators see national governments as main focus for action. NGOs and labour organisations focus more on international institutions and actions.*

Table 9.3   Summary of recommendations for action in response to the end of the MFA (private sector)

Global	*Garment exporting countries*
What can clothing brands and retailers do? **Possible actions:** - Source from countries that respect core labour standards and work with the government, suppliers, trade unions and NGOs to ensure decent working conditions. - Ensure that pricing practices are not at odds with labour standards compliance. - Work with suppliers to ensure they have reserve funds to address retrenchment of workers or closedown of facility in a responsible manner and in line with national law. Give adequate notice for ending supplier relationships. - Support government efforts in retraining and job banks for workers—both within the industry and for retrenched workers. - Support government efforts to upgrade the industry. - Ensure suppliers pay legally required social security/pension payments.	**What can clothing manufacturers do?** **Possible actions:** - Upgrade technology, management and skills of workers in order to remain competitive and diversify into higher-value-added garments and processes. - Improve factory standards and working conditions in order to improve quality and ensure market access. - Lobby and work with government and others to develop national industrial policies to which support real competitiveness. - Ensure workers have access to job banks and retraining programmes. - Ensure workers are paid their rights in retrenchment according to the law—don't just 'cut and run.' - Pay legally required social security/pension payments and ensure they are up to date. - Ensure their contractors and subcontractors are compliant.
Agreements: *Predominant recommendations in this area focus on mitigating negative impacts of changes in trade, not on effecting volumes of production and trade.*	**Agreements:** *Many manufacturers are international in operation and are not wedded to the countries in which they currently operate. It is not just orders that are leaving countries, but whole factory operations.*
Disagreements: *The key disagreement here is between those NGOs and trades unions that say that buying companies should remain loyal to existing suppliers and other organisations that do not see this as part of their responsibility.*	**Disagreements:** *The basic disagreement is between those factories that believe cheap labour is vital to remain competitive and those who do not. There is an underlying difference of perception: are workers considered disposable tools to be exploited or capital asset to be invested in through training and development?*

Source: Accountability (2004)

mean the development of responsible exit strategies, such as giving an appropriate period of notice to local manufacturers and establishing funds to ensure that workers are paid their redundancy entitlements. It would also involve monitoring the impact of quota removal on workers and communities and mobilising companies to work together with governments and local organisations to take remedial action, such as the retraining of workers. Beyond this, it is seen as part of the remit of the Forum to work towards agreement on how to achieve basic standards of good practice in the post-MFA restructuring of the industry. It has yet to be seen how effective this will be. However, it is clear that this initiative has provided trade unions and NGOs with a framework for working not only with sympathetic companies but also with governments and international institutions in a way that provides the capacity to bring about real changes. A platform now exists which provides the opportunity to establish a relationship between successful business practice and decent working conditions.

Conclusion

The phase-out of the MFA has been presented as a progressive measure that will benefit developing countries by opening up Northern markets. In reality, the main impact will not be a relocation of the industry from industrialised countries of the North to poorer developing countries in the South, but a relocation of production sites from one developing country to another. In particular, there is likely to be an overall shift of production to China, with Bangladesh predicted to be the main loser. The main beneficiaries will be Northern-based companies that will profit from fewer restrictions on their global operations. Workers in both industrialised and developing countries will suffer the consequences. Not only will there be massive job losses but, even in countries which benefit from the phase-out, the downward pressure on wages and working conditions is likely to become even more intense.

The phase-out of the MFA is combining with the expansion of subcontracting to cause a crisis in the garment industry from the perspective of workers. Many are facing increased insecurity and deteriorating terms of employment as the industry becomes more and more dependent on the exploitation of inequality both within and between nations. However, there are signs that this is not an irreversible trend. It is predicted that the consolidation of the industry associated with the MFA phase-out will take place at a national as well as an international level, which could mean that, in the long run, more workers become employed in larger, regulated factories. At the same time, the current deterioration of labour standards is being increasingly recognised as unacceptable, with big clothing companies responding to the demand that they take responsibility for the implications of their business practices on workers. Some companies are beginning to realise that it is in their interests to work together with trade unions and NGOs to confront the negative implications of restructuring, not only in the countries where they are based, but in the world as a whole.

10

Conclusion

Angela Hale with Jane Wills

In this book we have attempted to develop new insights into the global garment industry by drawing together the findings of action research recently carried out by members of the Women Working Worldwide network and associated organisations. The focus has been on how the changing structure of the industry, and in particular the expansion of subcontracting, has impacted on the lives of workers and how new strategies of resistance have developed in response to the negative aspects of this impact. These strategies have involved the development of local and global networks involving both workers and consumers and engagement with the Northern-based corporate social responsibility initiatives that have emerged in response to public exposure of workers' rights abuses. Taken together these organisational responses can be seen as themselves contributing to the continual reshaping of the industry.

Currently, the garment industry operates as one of the clearest examples of the mode of capitalist production associated with neo-liberal globalisation, a key feature of which is competitive sourcing involving the exploitation of inequality at both national and international level. This process structures the industry both horizontally, through market competition between nations and firms, and vertically, through the process of subcontracting. The research reported in this volume provides new insights into how this competitive subcontracting has expanded the reach of garment industry supply chains both geographically and socially. Most who work within the industry are themselves unaware of the extent of this reach, let alone the conditions under which workers are employed at the far end of supply chains.

The hidden nature of subcontracting chains serves to obscure the extent and depth of worker exploitation. Research in all locations clearly revealed that the abuse of workers' rights is greatest the furthest away they are from the centres of production. Workers at the end of the supply chain are facing lower wages, poorer conditions and greater insecurity than those in factories further up the chain. It also revealed how subcontracting and the informalisation of employment relations takes advantage of social inequalities relating not only to levels of poverty but also to gender, age, ethnicity and migration. This was found to be the case even in locations in Mexico where 'full-package production' is portrayed as a mechanism for upgrading the local industry (see Chapter 7). Any approach to improving working conditions in the industry therefore needs to confront the significance of subcontracting for workers' rights and explore how workers in subcontracted units can organise in the face of strong downward pressures.

Traditionally, such challenges have been tackled through collective organisation, backed up by state legislation and institutional interventions by national- and international-level bodies such as fair wages boards, training bodies and the International Labour Organisation. However, our research confirms that trade unions only have an active presence in a limited number of workplaces towards the top end of the chain. Furthermore, international competition for capital investment and trade has made governments less willing to intervene in order to protect and improve conditions of work. In this political vacuum, a number of new organisations have emerged to support, organise and advocate on behalf of those involved in the production of goods for export in the developing world. These organisations have been particularly significant in political environments where trade union organisation has not been possible, for example in EPZs. Many have been set up specifically to support women workers and as such have developed creative forms of resistance, using spaces outside the workplace to reach workers through their communities, making explicit connections between home and work. As WWW we have sought to develop long-term links with these organisations and to support their priorities rather than impose any models from the outside. This is how the action research reported in *Threads of Labour* arose: organisations in WWW's network needed to know more about the way in which corporations are structured and what this means for their work and political organisation (see Chapters 3 and 4).

The research reported in this volume is embedded in attempts by labour activists to face up to the challenges presented by the complex and shifting nature of the globalised garment industry. Carrying out this action research as a series of locally rooted but internationally co-ordinated projects has highlighted the commonalities and differences between places involved in the global garment industry. The complexities revealed have demonstrated that forging international solidarity is anything but straightforward. As eloquently stated by Rohini Hensmann in Chapter 8, workers are differentially positioned in supply chains on the basis of geography, corporate hierarchy and workplace status. Even within any one country and any one factory, different garment workers will have more or less power to act in order to improve their conditions of work. The strategies available to some will not be within reach of, or applicable to, others. And, in this context, we argue that solidarity is best crafted by taking a lead from the workers producing the goods, by working with organisations that are trying to solve workers' own problems, whether or not these are trade unions in the traditional sense. As WWW we have been able to build on such solidarity to also inform our involvement in Northern-based corporate social responsibility initiatives and to contribute to debates about reshaping the structure of the industry.

The work of WWW can be seen as part of a new form of industrial action which involves political alliances between workers, trade unions, local and regional activists and consumer-based organisations in the key markets and central locations of major buyers (see Chapter 3). The significance of these alliances for bringing about change is demonstrated by the evidence we have provided of cases in which there have been notable improvements in working conditions. Yantz (Chapter 7), for example, notes that the most widespread improvements in working practices have been in matters such as child labour and pregnancy testing, which have been the focus of Northern-based campaigns. Hensman (Chapter 8) and Hale (Chapter 3) similarly provide examples of how internationally co-ordinated campaigns have successfully contributed to the establishment of trade union rights in a number of different locations. And, significantly in the case of workers in Bombay/Mumbai, how their support for workers in the USA helped those Northern workers win a trade union recognition dispute (see Chapter 8). However, these successes almost always apply to workplaces relatively high up subcontracting chains. The likelihood of successful international campaigns in support of workers rights in small informalised workplaces is extremely limited, particularly if those workplaces are supplying numerous buyers further up corporate chains.

For such workers, the most useful strategy may be lobbying national governments to establish full legal rights for all workers. Alliances with international campaigning organisations could assist in such work by making demands for the formalisation of employment relationships at all levels, and by putting pressure on bodies like the ILO, the governments of Northern consuming nations and multi-stakeholder bodies to work towards this. Unless workers have proof of employment, there is little hope of their being able to use codes of conduct, ILO conventions or consumer pressure as negotiating tools. Indeed, it is becoming clear that campaigning bodies at all levels need to address the structural injustices of the industry, and particularly the impact of subcontracting, on a more systematic basis. As the extent and impact of subcontracting is becoming more apparent, it is critically important to address the way in which this erodes terms and conditions down corporate chains.

In this regard, it is important to ensure that informalisation is addressed by the multi-stakeholder initiatives that have been established in response to international campaigns. Companies signing up to the UK Ethical Trading Initiative, for example, are now clear that their responsibility is to work towards implementation of the base code of good practice down their supply chains to homeworkers. The huge challenge that this presents has encouraged some of the more responsible companies to work towards the greater consolidation of their supply chains. As Hensman argues, this trend may initially lead to job losses for some workers, but it will increase the overall proportion of workers employed in more regulated workplaces in the long term. A similar trend may emerge as an outcome of the MFA phase-out (see Chapter 9): while thousands of workers are expected to lose their jobs as a result of global relocation, some experts are predicting that a decrease in geographical dispersion will be accompanied by a move towards larger production units and the chance to improve conditions of work.

If such restructuring occurs, it will result more from corporate self-interest than any local or international labour campaigns. But it is now becoming clear that it is in the self-interest of many major retailers and brand merchandisers to incorporate appropriate responses to the demands from both workers and consumers to address the abuse of workers' rights in the industry. This is clearly demonstrated in the new multi-stakeholder initiative that has developed in response to the likely impact on workers of the MFA phase-out reported in Chapter 9. It is now unacceptable for major players in the industry to ignore the plight of the workers in their supply chains, and they are increasingly receptive to information from both worker and consumer-based organisations. This shift in attitude can be seen as a major achievement for the many organisations championing the rights of garment workers. Resistance to the exploitative aspects of the garment industry can now be seen as contributing to the restructuring of the industry on a more humane basis.

Although *Threads of Labour* is focused solely on the garment industry, the research and action reported in this volume can also be seen as relevant to those tackling the economic and social injustices of production in other economic sectors, as well as what has come to be called the wider global justice movement. Work on labour conditions in other consumer supply chains, such as horticulture and electronics, has revealed similar downward pressures from retailers, the restructuring of production chains and the associated threats to the rights of mainly women workers (Barrientos et al, 2003; Ferus-Comelo, 2005; Hale and Opondo, 2005; Raworth, 2004). The contributors to this volume have demonstrated that such complex supply chains require a multi-scalar and multi-actor approach to improving working conditions. The activities that we have developed as WWW involve international networks between local organisations that support, organise and represent workers in the garment industry, along with activist groups based in consuming nations. The composition of these networks varies depending on the issue being considered and the project being completed. They involve trade unions working alongside more community-focused organisations; groups based on class identities (focused on workplace interests) together with those mobilised around gender, religion, ethnicity and geography; those with traditional organisational hierarchies and those with none. On a small scale, these activities highlight the importance of what Routledge (2003; see also Cumbers and Routledge, 2004) has called 'convergence spaces' that are created by transnational networks resisting aspects of neo-liberal globalisation. He points to the way in which such networks engage in communication, information sharing, solidarity, co-ordination and resource mobilisation. These practical activities have always been at the heart of our work to resist the impact of global production as WWW, and they suggest ways in which the international labour movement could be reinvigorated in its creative encounter with capital.

It is increasingly apparent in an age of globalisation that questions of labour rights and employment standards cannot be tackled by the trade union movement acting alone. Working conditions, and particularly wage rates, are of critical importance in determining the economic, political and social outcomes of industrialisation and globalisation, and hence, the very process of international development. Labour standards matter to the well-being of whole nations and communities, as well as individual workers. As such, it is possible to construct a very broad coalition of transnational actors and institutions to endorse greater economic and social justice for workers and their communities in the South. Moreover, groups like Jubilee 2000, the Trade Justice Movement, the Clean Clothes Campaign and elements of the wider global justice movement have highlighted the extent to which such coalitions can work to make significant gains. This approach involves linking actors and organisations that have shared goals, rather than common ideology and/or organisational structure.

In this volume we have attempted to show how, as WWW, we have networked with organisations on the basis of a common concern about the rights of garment workers and how, through this, a mixed coalition of workers, activists and consumers has found ways to act together, using different tactics at different times and on different scales (including a mixture of action research, education, information exchange, organising and advocacy). In presenting an analysis of the complexities of the garment industry alongside strategies of resistance being developed to ameliorate the impact on workers, *Threads of labour* bears testament to both the scale of the challenges faced and the potential of international networks to respond to those challenges in the years ahead.

References

Accountability (2004) *Managing the Transition to a Responsible Post MFA Global Garment Industry.* London: Accountability

Adams R L (2002) Retail profitability and sweatshops: A global dilemma. *Journal of Retailing and Consumer Services* 9:147–153

Alexander R and La Botz D (2003) Mexico's labor law reform. *Mexican Labor News and Analysis.* 8(April):4

Anson R (2003) The 'Big Bang': Winners and losers in the textile and clothing industry in 2005 and beyond. *Textiles Intelligence* 107(September–October):3–5

Appelbaum R (2003) *Assessing the Impact of the Phasing-out of the Agreement on Textiles and Clothing on Apparel Exports on the Least Developed and Developing Countries.* Santa Barbara: Center for Global Studies Institute for Social, Behavioral, and Economic Research, UC Santa Barbara

Ayuntamiento Municipal de Tehuacan (AMT) (2002) *Portafolios Informativo de Tehuacan: Tehuacan en Cifras, 1999–2002.* Tehuacan: AMT

Bair J and Gereffi G (2001) Local clusters in global chains: The causes and consequences of export dynamism in Torreon's blue jeans industry. *World Development* 29:1885–1903

Banerjee N (1981) Is small beautiful? In A K Bagchi and N Banerjee (eds) *Change and Choice in Indian Industry.* Calcutta: KP Bagchi

Barrie L (2003) 'Private Label: The State of the Market.' *Just-Style.com.* HTML file: <URL: www.just-style.com/membersclub/briefings.asp>, last accessed October 2004

Barrientos S (2001) Gender flexibility and global value chains. *IDS Bulletin* 32(3):83–94

Barrientos S, Dolan C and Tallontyre A (2003) A gendered value chain approach to codes of conduct in African horticulture. *World Development* 31:1511–26

Barrios M and Hernández R (2004) Foreword to second printing: Comisión de Derechos Humanos y Laborales del Valle de Tehuacan, A.C., en colaboración con la Red de Solidaridad de la Maquila (Canadá). In M Barrios and R Hernández *Tehuacán: Del Calzón de Manta a los Blue Jeans* (pp iii–vii). Toronto: Maquila Solidarity Network

Basatia B K, Kaur S and Canaan J E (1999) *The 'Double Burden' Intensified: Asian Women's Perceptions of Homeworking.* Birmingham: AEKTA Project

Baughman L, Mirus R, Morkre M, and Spinager D (1997) Of tyre cords, ties and tents: Window dressing in the ATC. *World Economy* 20:407–437

Begg B, Pickles J and Smith A (2003) Cutting it: European integration, trade regimes and the reconfiguration of East-Central European apparel production. *Environment and Planning A* 35:2191–2207

Bonacich E (2000) Intense challenges, tentative possibilities: Organizing immigrant garment workers in Los Angeles. In R Milkman (ed) *Organizing Immigrants: The Challenge for Unions in Contemporary California* (pp 130–149). Ithaca: ILR Press/Cornell University Press

Bradbury H and Reason P (2001) Conclusion: Broadening the bandwidth of validity: issues and choices—points for improving the quality of action research. In P Reason and H Bradbury (eds) *Handbook of Action Research: Participative Enquiry and Practice* (pp 447–455). London: Sage

Breitbart M (2003) Participatory research methods. In N Clifford and G Valentine (eds) *Key Methods in Geography* (pp 161–178). London: Sage

Bronfenbrenner K (2000) *Uneasy Terrain: The Impact of Capital Mobility on Workers, Wages and Union Organizing.* Report submitted to the US Trade Deficit Review Commission, available from the author, New York State School of Industrial and Labor Relations, Cornell University, USA

CAPITB Trust (2002) *Skills for Success: A report on the Impact of Market Changes on the Skill Base of the UK Clothing Industry.* Leeds: CAPITB Trust

Carnegie Council on Ethics and International Affairs (CCEIA) (2000) Who can protect workers' rights? The workplace codes of conduct debates. *Human Rights Dialogue* 2(4)

Central American Women's Network (CAWN) (1999) *Women Workers and Codes of Conduct: Workshop Report.* Manchester: WWW

Chambers R (1994a) Participatory Rural Appraisal (PRA): Analysis of experience. *World Development* 22(9):1253–1268

Chambers R (1994b) Participatory Rural Appraisal (PRA): Challenges, potentials and paradigm. *World Development* 22(10):1433–1454

China Labour Education and Information Centre (CLEIC) (1996) *The Flip Side of Success: The Situation of Workers and Organising in Foreign-invested Enterprises in Guangdong.* Hong Kong: CLEIC

Christerson B and Appelbaum R (1995) Global and local subcontracting: Space, ethnicity, and the organisation of apparel production. *World Development* 8:1363–1374

Clean Clothes Campaign (CCC) (1999) *Made in Eastern Europe.* Amsterdam: CCC

Clean Clothes Campaign (CCC) (2001) 'Campaign to Support the Free Trade Zone Workers Union of Sri Lanka.' HTML file: <URL: www.cleanclothes. org/urgent/01-09-23.htm>, last accessed October 2004

Clean Clothes Campaign (CCC) (2003) *CCC Newsletter* 17 (December). Amsterdam: CCC

Cleveland S H (2003) Why international labor standards? In R J Flanagan and W B Gould IV (eds) *International Labor Standards: Globalization, Trade, and Public Policy* (pp 128–178). Stanford: Stanford University Press

Clothesource Limited (2003) *PriceTrak EU & US Q1 2003.* Charlebury: Clothesource

Coats (2002) *Annual Report 2001.* Uxbridge: Coats Plc

Committee for Asian Women (CAW) (2002) *Moving Mountains: Twenty-five Years of Perseverance.* Bangkok: CAW

Cook I (2004) Follow the thing. *Antipode* 36:642–664

Cornwall A and Jenkes R (1995) What is participative research? *Social Science and Medicine* 41(12):1667–1676

Cravey A (2004) Students and the anti-sweatshop movement. *Antipode* 36:203–208

Cumbers A and Routledge P (2004) Alternative geographical imaginations: Introduction. *Antipode* 36:818–828.

Da Bindu (1990) *Newsletter.* Colombo: Da Bindu

Da Bindu (2000) *Strong Daughters: A Study of Women Workers in the Free Trade Zones of Sri Lanka.* Manchester: WWW

Dannecker P (2000) Collective action, organisation building, and leadership: Women workers in the garment sector in Bangladesh. *Gender and Development* 8:31–39

David Rigby Associates (2002) *North West Textile Sector.* Manchester: NWDA

De Coster J (2004) 'Cambodia Gets a Boost from Social Responsibility.' *Just-Style.com.* HTML file: <URL: www.just-style.com/features_detail. asp?art=729>, last accessed October 2004

De Haan E and Phillips G (2002) *Made in Southern Africa.* Amsterdam: Clean Clothes Campaign

De Jonquières G (2004) 'Garment Industry Faces a Global Shake Up.' *Financial Times* 19 July. HTML file: <URL yaleglobal.yale.edu/display.article>, last accessed November 2004

Development in Practice (2004) Special issue on unions and NGOs. 14(1–2).

Dhanarajan S (2004) *Play Fair at the Olympics: Bringing Justice to Workers in the Sportswear Industry.* Oxford, Amsterdam and Brussels: Oxfam International, Clean Clothes Campaign, International Confederation of Free Trade Unions

Diao X and Somwaru A. (2002) A global perspective of liberalising world textile and apparel trade. *Nordic Journal of Political Economy* 28(2):127–146

Dicken P (2003) *Global Shift: The Internationalization of Economic Activity* (3rd edition). London: Sage

Dicken P (2004) Geographers and globalisation: (Yet) another missed boat? *Transactions of the Institute of British Geographers* 29:5–26

Dicken P and Hassler M (2000) Organizing the Indonesian clothing industry in the global economy: The role of business networks. *Environment and Planning* 32:263–280

Dicken P, Kelly P F, Olds K and Wai-Chung Yeung H (2001) Chains and networks, territories and scales: Towards a relational framework for analysing the global economy. *Global Networks* 1:89–112

Diller J (1999) A social conscience in the global marketplace? Labour dimensions of codes of conduct, social labelling and investor initiatives. *International Labour Review* 138:99–130

Donald A and Bamford T (1990) Don't wear it. *International Labour Reports* 38:5–7

Dutch Trade Union Federation (FNV) (2003) *From Marginal Work to Core Business: European Trade Unions Organising in the Informal Economy.* Amsterdam: FNV

East Midlands Development Agency (EMDA) (2001) *Developing the Clothing and Textile Cluster in the East Midlands.* Nottingham: EMDA

El Sol de Tehuacan (2001) 4 October

Ellison L (1999) *Monitoring the Introduction and Impact of the National Minimum Wage on Homeworkers.* Leeds: National Group on Homeworkers

Elson D and Pearson R (1981) The subordination of women and the internationalisation of factory production. In K Young, C Wolkowitz and R McCullagh (eds) *Of Marriage and the Market: Women's Subordination in International Perspective* (pp 144–166). London: CSE Books

Elson D and Pearson R (1998) Nimble fingers revisited. In C Jackson and R Pearson (eds) *Feminist Visions of Development* (pp 171–176). London: Routledge

Enloe C (1990) *Bananas, Beaches and Bases: Making Feminist Sense of International Politics.* Berkeley: University of California Press

Ethical Trading Initiative (ETI) (2003) *Key Challenges in Ethical Trade: Report of the ETI Biennial Conference 2003.* London: ETI

Euratex (2004) *Bulletin 2004/1.* Brussels: Euratex

European Commission (2003) *Communication on the Future of the Textiles and Clothing Sector in the Enlarged European Union.* Final communication from the Commission to the Council, the European Parliament, the European Economic and Social committee and the committee of the regions, 29 October. Brussels: European Commission

Fair Labor Association (FLA) (2003) *First Public Report: Towards Improving Workers' Lives.* Washington: FLA

Ferus-Comelo A (2005) *Globalisation and labour organisation in the electronics industry.* Unpublished PhD thesis, University of London.

Fields G S (2003) International labor standards and decent work: Perspectives from the developing world. In R J Flanagan and W B Gould IV (eds) *Inter-*

national Labor Standards: Globalization, Trade, and Public Policy (pp 61–79). Stanford: Stanford University Press

Fisher W F and Ponniah T (eds) (2003) *Another World is Possible: Popular Alternatives to Globalization at the World Social Forum.* London: Zed

Flanagan M (2003) 'Apparel Sourcing in the 21st Century: The 10 Lessons So Far.' *Just-Style.com.* HTML file: <URL: www.just-style.com/membersclub/briefings.asp>, last accessed October 2004

Flanagan M and Leffman L (2001) Global apparel sourcing: options for the future. *Textile Outlook International* 94(July):129–169

Foo L J and Bas N F (2003) *Free Trade's Looming Threat to the World's Garment Workers.* Los Angeles: Sweatshopwatch

Fröbel F, Heinrichs J and Kreye O (1980) *The New International Division of Labour: Structural Unemployment in Industrialised Countries and Industrialisation in Developing Countries.* Cambridge: Cambridge University Press

Gereffi G (1994) The organization of buyer-driven global commodity chains: How US retailers shape overseas production networks. In G Gereffi and M Korzeniewicz, (eds) *Commodity Chains and Global Capitalism* (pp 95–122). Westport CT: Praeger

Gereffi G (1999) International trade and industrial upgrading in the apparel commodity chain. *Journal of International Economics* 48:37–70

Gereffi G (2001) Shifting governance structures in global commodity chains: With special reference to the Internet. *American Behavioural Scientist* 44:1616–1637

Gereffi G (ed) (2003) *Who Gets Ahead in the Global Economy? Industrial Upgrading, Theory and Practice.* Baltimore: Johns Hopkins Press

Gereffi G and Korzeniewicz M (eds) (1994) *Commodity Chains and Global Capitalism.* Westport CT: Praeger

Gereffi G and Martinez M (1999) 'Blue Jeans and Local Linkages: The Blue Jeans Boom in Torreón, Mexico.' *Background Paper for the 2000 / 2001 World Development Report on Poverty.* HTML file <URL www.worldbank.org/wbp/wdrpoverty/background/gereffi.pdf>, last accessed October 2004

Gereffi G, Humphrey J, Kaplinsky R and Sturgeon T J (2001) Introduction: Globalisation, value chains and development. *IDS Bulletin* 32(3):1–8

Gilbert H (2002) *Out of Sight … Out of Mind: A Report on the Health and Safety Needs of UK Homeworkers.* Leeds: National Group on Homeworking

Gopal J (ed) (2002) *Garment Industry in South Asia: Rags or Riches? Competitiveness, Productivity and Job Quality in the Post-MFA Environment.* New Delhi: International Labour Organization

Gordon J (2001) Organizing low wage immigrants: The workplace project—Interview with Jennifer Gordon. *Working USA* 5(1):87–102

Greenwood D and Levin M (1998) *Introduction to Action Research: Social Research for Social Change.* London: Sage

Guerra A (2002) *El Mundo de Tehuacan* 27 January

Hale A (2000a) What hope for 'ethical' trade in the globalised garment industry? *Antipode* 32:349–356

Hale A (2000b) *Phasing Out the Multi-Fibre Arrangement: What Does it Mean for Workers?* Manchester: WWW

Hale A (2002) Trade liberalisation in the garment industry: Who is really benefiting? *Development in Practice* 12(1):33–44

Hale A (2004) Beyond the barriers: New forms of labour internationalism. *Development in Practice* 14(1–2):158–161

Hale A and Opondo M (2005) Humanising the cut-flower chain: Confronting the realities of flower production for workers in Kenya. *Antipode* 36 (forthcoming)

Hammond J and Kohler K (2000) *E-Commerce in the Textile and Apparel Industries: Brookings Volume of BRIE Papers.* Boston: Harvard Business School

Harvey D (2000) *Spaces of Hope.* Edinburgh: Edinburgh University Press

Harvey M, Quilley S and Beynon H (2002) *Exploring the Tomato.* London: Edward Elger

Hassler M (2000) The Indonesian clothing industry: A commodity chain approach. In *WWW Seminar Report: Global Supply Chains, Gender and the Challenges to the Labour Movement: Grapes, Tomatoes, T-Shirts and Jeans* (pp 3–11). Manchester: WWW

Hayes S G and Jones R M (2002) The economic determinants of clothing consumption in the UK. *Journal of Fashion Marketing and Management* 6(4):326–339

Heintz J (2004) Beyond sweatshops: Employment, labor market security and global inequality. *Antipode* 36:222–226

Hensman R (2001) World trade and workers' rights: The search for an internationalist position. In P Waterman and J Wills (eds) *Place, Space, and the New Labour Internationalisms* (pp 123–146). Oxford: Blackwell Publishing

Heyzer N 1978 Young women and migrant workers in Singapore's labour intensive industries. Paper presented at the IDS Conference 'Continuing subordination of women in the development process.' Brighton: IDS

Holmes J (2004) Re-scaling collective bargaining: Union responses to restructuring in the North American auto industry. *Geoforum* 35:9–22

Holmstrom M (1984) *Industry and Inequality: The Social Anthropology of Indian Labour.* Cambridge, UK: Cambridge University Press

Homenet (1999) *The ILO Convention on Homeworking.* Leeds: Homenet

Homenet (2002) Organising for change. *Newsletter* (February). Leeds: Homenet

Howitt R (2002) Preface. In R Jenkins, R Pearson and G Seyfang (eds) *Corporate Responsibility and Labour Rights: Codes of Conduct in the Global Economy* (pp xiii–xvi). London: Earthscan

Hughes A (2001) Global commodity networks, ethical trade and governmentality: Organizing business responsibility in the Kenyan cut flower industry. *Transactions of the Institute of British Geographers* 26:390–406

Human Rights Watch (2004) CAFTA's weak labour rights protections: Why the present accord should be opposed. *Human Rights Watch Briefing Paper* (March):1

Humphrey J and Schmitz H (2001) Governance in global value chains. *IDS Bulletin* 32(3):19–29

Hurley J (2003) *Bridging the Gap: A Look at Gap's Supply Chain from the Workplace to the Store.* Manchester: WWW

Ianchovichina E and Martin W (2001) *Trade Liberalization in China's Accession to the WTO.* New York: World Bank

International Confederation of Free Trade Unions (ICFTU) (1996) *Behind the Wire: Anti-union Repression in the Export Processing Zones.* Brussels: ICFTU

International Confederation of Free Trade Unions (ICFTU) (2003) *EPZs: Symbols of Exploitation and a Development Dead-End.* Brussels: ICFTU

International Labour Organisation (ILO) (2000) *Labour Practices in the Footwear, Leather, Textiles and Clothing Industries.* Geneva: ILO

International Labour Organisation (ILO) (2002) 'Resolution Concerning Decent Work and the Informal Economy'. *General Conference of the ILO 2002.* PDF file: <URL: www.ilo.org/public/english/standards/relm/ilc/ilc90/pdf/pr-25res.pdf>, last accessed November 2004

International Labour Organisation (ILO) (2003) ILO Database on export processing zones. Geneva: International Labour Office: PDF file <URL *www.ilo.org/public/english/dialogue/sector/themes/epz/epz-db.pdf*>, last accessed December 2004

Jassin-O'Rourke Group (2002) *Global Competitiveness Report: Selling to Full Package Providers.* New York: ILO

Jenkins R, Pearson R and Seyfang G (eds) (2002) *Corporate Responsibility and Labour Rights: Codes of Conduct in the Global Economy.* London: Earthscan

Johns R and Vural L (2000) Class, geography and the consumerist turn: UNITE and the stop the sweatshops campaign. *Environment Planning A* 32:1193–1214

Jones R M (2002) *The Apparel Industry.* London: Blackwell

Junta Local de Conciliación de Tehuacan (JLCT) (2002) *Informe Estadístico.* Tehuacan: JLCT

Kabeer N (2000) *The Power to Choose: Bangladeshi Women and Labour Market Decisions in London and Dhaka.* London: Verso

Kabeer N (2004) Globalization, labor standards and women's rights: Dilemmas of collective (in)action in an interdependent world. *Feminist Economics* 10:3–35

Kaplinsky R (2000) Spreading the gains from globalisation: What can be learned from value chain analysis? *Journal of Development Studies* 37:117–146

Kearney N (2003a) 'What Future for Textiles and Clothing Trade after 2005?' *ITGLWF Press Release* 2 September. HTML file: <URL: www.itglwf.orrg/

displaydocument.asp?DocType=Press&Language=&Index=595>, last accessed October 2004

Kearney N (2003b) 'WTO Urged to Right Damage it has Done to the Clothing and Textile Industry.' *ITGLWF Press Release* 10 September. HTML file: <URL: www.itglwf.org/displaydocument.asp?DocType=Press&Index=Language=EN>, last accessed November 2004

Kelegama, S and Epaarachchi, R (2001) 'Productivity, Competitiveness and Job Quality in Garment Industry in Sri Lanka.' Discussion paper for the ILO.

KFAT News (1999) 'Manufacturing our Future' Summer:8

Kibria N (1995) Culture, social class and income control in the lives of women garment workers in Bangladesh. *Gender and Society* 9(3):289–309

Klein N (2000) *No Logo: Taking Aim at the Brand Bullies*. London: Flamingo

Krishnamoorthy S (2004) *Structure of the Garment Industry and Labour Rights in India: The Post MFA Context*. New Delhi: Centre for Education and Communication

La Journada de Oriente (2004) Ante las presiones, el gobernador niega los registros para los sindicatos del Hospital del Niño Poblano. June 10

Labour Behind the Label (LBL) (2001) *Wearing Thin: The State of Pay in the Fashion Industry*. Norwich: LBL

Lips M, Tabeau A, van Tongeren F, Ahmed N and Herok C (2003) Textile and wearing apparel sector liberalization: Consequences for the Bangladesh economy. Paper presented at the 'Conference on Global Economic.' The Hague, June 12–14

Lund F and Nicholson J (eds) (2004) *Chains of Production, Ladders of Protection: Social Protection for Workers in the Informal Economy*. Washington: World Bank

Lyne N (2002) Profile of Inditex: Building on the success of Zara. *Textile Outlook International* 99(May–June):166–181

Magretta J (1998) Fast, global and entrepreneurial: Supply chain management Hong Kong style: An interview with Victor Fung. *Harvard Business Review* (September–October):103–114

Majmudar M (1996) Trade liberalisation in clothing: The MFA phase out and developing countries. *Development Policy Review* 14:5–36

Maquila Solidarity Network (MSN) (2003a) *Tehuacan: Blue Jeans, Blue Waters and Worker Rights*. Toronto: MSN

Maquila Solidarity Network (MSN) (2003b) Thai Workers win settlement, no thanks to La Senza or Jacob. *Maquila Network Update* 8(3):3–8

Maquila Solidarity Network (MSN) (2003c) *Codes Memo* (September). Toronto: MSN

Marx K (1977) *Capital*. New York: Random House

Massey D (2004) Geographies of responsibility. *Geografiska Annaler* 86B:5–18

McCormick D and Schmitz H (2002) *Manual for Value Chain Research on Homeworking in the Garment Industry*. Brighton: IDS

McDowell L (1992) Doing gender: Feminism, feminists and research methods in human geography. *Transactions of the Institute of British Geographers* 16:400–419

Mendez A (2001) Denuncian a 100 maquiladoras clandestinas, trabajan sin permiso y no dan prestaciones a trabajadores. *El Mundo de Tehuacan* 21 January

Mehta P (1997) Textiles and clothing: Who gains, who looses and why! *Working Paper No 5*. Calcutta: Centre for International Trade, Economics and Environment

Mexican Action Network on Free Trade (RMALC) 'NAFTA and Labor Conditions in Mexico.' HTML file: <URL: www.developmentgap.org/rmalclab.html>, last accessed October 2004

Miller D (2004) Preparing for the long haul: Negotiating international framework agreements in the global textile, garments and footwear sector. *Global Social Policy* 4(2):215–239

Miller D and Grinter S (2003) International framework agreements in the global textile, garment and footwear sector. *Managerial Law* 45:111–116

Ministry of Labour (2002) *Report of the National Commission on Labour.* New Delhi: Government of India

Mollet A (2001) Profile of Ramatex: A Malaysian group with investments in China and South Africa. *Textile Outlook International* (May–June): 89–114

Monbiot G (2003) *The Age of Consent: A Manifesto for a New World Order.* London: Flamingo

Municipality of Tehuacan (2002) *Municipal Economic Census.* Tehuacan: Municipality of Tehuacan

Munck R (2002) *Globalisation and Labour.* London: Zed Books

Musiolek B (2004) 'Made in Eastern Europe: The New Fashion Colonies.' HTML file: <URL: www.cleanclothes.org/publications/04-01-made-in-eastern-europe.htm>, last accessed November 2004

Namibian (2004) 'Chinese Leave Ramatex in Droves.' *www.namibian.com* 4 February. HTML file: <URL: www.namibian.com.na/2004/february/national/0421CE2C93.html>, last accessed October 2004

National Interfaith Committee for Workers' Justice (NICWJ) (1998) *Cross Border Blues: A Call for Justice for Maquiladora Workers in Tehuacan.* Chicago: NICWJ

Ness I (1998) Organizing immigrant communities: UNITE's workers' centre strategy. In K Bronfenbrenner, S Friedman, R W Hurd, R Oswald and R L Seeber (eds) *Organizing to Win: New Research on Union Strategies* (pp 87–101). Ithaca: ILR Press/Cornell University Press

Nordas H K (2004) *The Global Textile and Clothing Industry Post the Agreement on Textile and Clothing.* Geneva: WTO

Nyden P, Figert A, Shibley M and Burrows D (1997) *Building Community: Social Science in Action.* Thousand Oaks, CA: Pine Forge Press

Office of National Statistics (ONS) (2002) Earnings and employment. *Labour Market Trends* (July). London: ONS

Ofreneo R P, Lim J Y and Gula L A (2001) The view from below: Impact of the financial crisis on subcontracted workers in the Philippines. In R Balakrishnan (ed) *The Hidden Assembly Line: Gender Dynamics of Subcontracted Work in a Globalised Economy* (pp 87–114). Bloomfield: Kumarian Press

Osorio Y B (2002) Sancionarian con 250 mil *pesos* a Cualquier Lavado. *El Mundo de Tehuacan* 28 November

Oxborrow L (1999) 'Beyond Needles and Thread: Changing Supply Chains in the UK.' *The Harvard Center for Textile and Apparel Research.* PDF file: <URL www.hctar.org/pages/pubs.html>, last accessed November 2004

Oxborrow L (2002) *The EU Textiles Sector in 2008: Up-skilling for the Future.* Lecture delivered at the Nottinghamshire International Clothing Centre

Oxfam International (2004) *Stitched Up: How Rich-Country Protectionism in Textiles and Clothing Trade Prevents Poverty Alleviation.* London: Oxfam International

Paredes C J (1910) *Apuntes Historicos de Tehuacan.* Tehuacan: Paredes Colin

Parmar C and Purdey K (2000) The implications of increased international subcontracting for UK workers. In H Plunkett (ed) *Organising Along International Subcontracting Chains in the Garment Industry: Women Working Worldwide Workshop.* Manchester: Women Working Worldwide

Payne M (2002) Profile of Levi Strauss. *Textile Outlook International* 97(January–February):9–31

Pearson R (1994) Organising home-based workers in the global economy: An action-research approach. *Development in Practice* 14:136–148

Perez Cote H (2001) *El Mundo de Tehuacan* 2 October

Perrons D (2004) *Globalization and Social Change: People and Places in a Divided World.* London: Routledge

Pollin R and Luce S (1998) *The Living Wage: Building a Fair Economy.* New York: New Press

Popp A D, Ruckman J E and Roew H D (2000) Quality in international clothing supply chains: UK companies' perspectives *Journal of Fashion Marketing and Management* 4(4): 351–360

Prieto M and Quinteros C (2004) Never the twain shall meet? Women's organisations and trade unions in the maquila industry in Central America. *Development in Practice* 14(1–2): 149–157

Ramírez Cueva J (2001) Tehuacan: La capital de los jeans. *La Jornada* 29 July

Raworth K (2004) *Trading Away Our Rights: Women Working in Global Supply Chains.* Oxford: Oxfam International

Reason P and Bradbury H (eds) (2001) *Handbook of Action Research: Participative Enquiry and Practice.* London: Sage

Retail Forward Inc (2003) 'Wal-Mart Apparel: Beyond Basics.' *www.retailforward.com.* HTML file: <URL: www.retailwire.com/RetailForward/Wal-Mart_Apparel.fm>, last accessed October 2004

Reyes J (1997) Gumaro Amaro: 8 anos de silencio. *Sintesis* 17 February

Reynolds D (2001) Living wage campaigns as social movements: Experiences from nine cities. *Labor Studies Journal* 26(Summer):31–65

Rock M T (2003) Public disclosure of the sweatshop practices of American multinational garment/shoe makers/retailers: Impacts on their stock prices. *Competition and Change* 7:23–38

Rosa K (1983) *Testimonies of Women Garment Workers* (Unpublished document). WWW: Manchester

Rosa K (1994) The conditions and organizational activities of women in Free Trade Zones: Malaysia, Philippines and Sri Lanka, 1970–1990. In S Rowbotham and S Mitter (eds) *Dignity and Daily Bread: New Forms of Economic Organizing among Poor Women in the Third World and the First* (pp 73–99). London: Routledge

Ross A (ed) (1997) *No Sweat: Fashion, Free Trade and the Rights of Garment Workers*. London: Verso

Routledge P (2003) Convergence space: process geographies of grassroots globalization networks. *Transactions of the Institute of British Geographers* 28: 333–49.

Rowbotham S and Mitter S (eds) (1994) *Dignity and Daily Bread: New Forms of Economic Organising Among Poor Women in the Third World and the First*. London: Routledge

Shaw L (1997) European Clean Clothes Campaigns. In A Ross (ed) *No Sweat* (pp 215–220). New York: Verso

Shaw L (2002) *Refashioning Resistance: Women Workers Organising in the Global Garment Industry*. Manchester: WWW

Shaw L and Hale A (2002) The emperor's new clothes: What codes mean for workers in the garment industry. In R Jenkins, R Pearson and G Seyfang (eds) *Corporate Responsibility and Labour Rights: Codes of Conduct in the Global Economy* (pp 101–112). London, Earthscan

Shepherd E (2001) 'SA8000: An External Code of Conduct, Accreditation and Monitoring System'. *Asian Labour Update* 39(April–June). HTM file: <URL: www.amrc.org.hk/Arch/3706.htm>, last accessed October 2004

Sinha S (1996) Child labour and education policy in India. *The Administrator* XLI (July–September):17–29

Smit M (1989) *C&A: The Still Giant*. Amsterdam: SOMO

Smith A (2003) Power relations, industrial clusters and regional transformations: Pan-European integration and outward processing in the Slovak clothing industry. *Economic Geography* 79:17–40

Smith A, Rainnie A, Dunford M, Hardy J, Hudson R and Sadler D (2002) Networks of value, commodities and regions: Reworking divisions of labour in macro-regional economies. *Progress in Human Geography* 26:41–63

Someya M, Shunnar H, and Srinvasan T G (2002) *Textile and Clothing Exports in MENA: Past Performance, Prospects and Policy Issues in Post MFA Context*. World Bank: Washington

Standing G (1989) Global feminisation through flexible labour: A theme revisited. *World Development* 27:583–602

State Centre for Municipal Development (SCMD) (1999) *Programa Emergente de Salud Para Empresas Maquilas*. Tehuacan: SCMD

Sturgeon T (2001) How do we define value chains and production networks? *IDS Bulletin* 32(3):9–18

Sweatshopwatch (2003) 'The Power of Retailers and Their Private Labels.' HTML file: <URL: www.sweatshopwatch.org/swatch/industry/cal/retailers/html>, last accessed October 2004

Teran Soto R (2002) Mueren intoxicados 2 obreros. *El Mundo de Tehuacan* 28 November

Textile and Clothing Strategy Group (TCSG) (2002) *A National Strategy for the UK Textile and Clothing Industry: Making It Happen* London: Department of Trade and Industry

Textile Outlook International (1996) Liberalisation of world trade in textiles and clothing: The views of exporting and importing countries. *Textile Outlook International* 66(July)

Tirado S (1994) Weaving dreams, constructing realities: The nineteenth of September National Union of Garment Workers in Mexico. In S Rowbotham and S Mitter (eds) *Dignity and Daily Bread: New Forms of Economic Organizing among Poor Women in the Third World and the First* (pp 100–113). London: Routledge

Transnational Information Exchange-Asia (TIE-Asia) (2004) HTML file: <URL: www.tieasia.org/SL.htm>, last accessed October 2004

Traub-Werner M and Cravey A (2002) Spatiality, sweatshops and solidarity in Guatemala. *Social and Cultural Geography* 3:383–401

Travis L E and Greene F J (2000) *Working with the Euro: Jo-y-Jo* Durham: Durham University Business School

Tyler D (2001) *Strategic Direction for Textiles and Clothing in Rochdale*. Manchester: North West Advanced Clothing Web

Underhill G (1998) *Industrial Crisis and the Open Economy: Politics, Global Trade and the Textile Industry in Advanced Economies*. London: Macmillan

Unni J and Bali N (2002) Subcontracted women workers in the garment industry in India. In R Balakrishnan (ed) *The Hidden Assembly Line: Gender Dynamics of Subcontracted Work in a Globalised Economy* (pp 115–144). Bloomfield: Kumarian Press

Wainwright H (2003) *Reclaim the State*. London: Verso

Walkenhorst P (2003) *Liberalising Trade in Textiles and Clothing: A Survey of Quantitative Studies*. Paris: OECD

Walsh J (2000) Organizing the scale of labour regulation in the United States: Service sector activity activism in the city. *Environment and Planning A* 32:1593–1610

Wills J (2001a) Community unionism and trade union renewal in the UK: Moving beyond the fragments at last? *Transactions of the Institute of British Geographers* 26:465–483

Wills J (2001b) *Mapping Low Pay in East London.* Report prepared for TELCO's Living Wage Campaign. London: TELCO

Wills J (2003a) Bargaining for the space to organise in the global economy: A review of the Accor–IUF trade union rights agreement. *Review of International Political Economy* 9:675–700

Wills J (2003b) *On the Front Line of Care: A Research Report to Explore Home Care Employment and Service Provision in Tower Hamlets.* London: UNISON

Wills J (2004) Organizing the low paid: East London's living wage campaign as a vehicle for change. In G Healy, E Heery, P Taylor and W Brown (eds) *The Future of Worker Representation* (pp 264–282). Basingstoke: Palgrave

Wills J et al (2002) *Report on the State of Civil Society in Plaistow and Canning Town, East London.* London: Queen Mary, University of London

Women in Informal Employment: Globalizing and Organizing (WIEGO) (2001) *Addressing Informality, Reducing Poverty: A Policy Response to the Informal Economy* Cambridge, MA: WIEGO

Women's International Network (WIN) (2002) Make yourself seen, make yourself count: International seminar on organising women in the informal sector. *WIN News* Spring

Women's Wear Daily (2004) 'Poland Ready for a Wild 2005' 13 July. HTML file: <URL: www.wwd.com>, last accessed November 2004

Women Working Worldwide (WWW) (1996) *World Trade is a Women's Issue: Conference Report.* Manchester: WWW

Women Working Worldwide (WWW) (1998) *Korean Outward Investment and the Rights of Women Workers: Conference Report.* Manchester: WWW

Women Working Worldwide (WWW) (1999) *Women Workers and Codes of Conduct: Asia Workshop Report.* Manchester: WWW

Women Working Worldwide (WWW) (2000a) *Globalisation and Informalisation: Report of International Workshop in Seoul.* Manchester: WWW

Women Working Worldwide (WWW) (2001) *Women Garment Workers Define their Rights: Reports from Sri Lanka, India and the Philippines.* Manchester: WWW

Women Working Worldwide (WWW) (2002a) *Company Codes of Conduct and Workers Rights: Report of an Education and Consultation Programme in Asia.* Manchester: WWW

Women Working Worldwide (WWW) (2002b) *Organising Along International Subcontracting Chains in the Garment Industry : Report of Workshop.* Manchester: WWW

Women Working Worldwide (WWW) (2003a) *Garment Industry Subcontracting and Workers Rights: Reports of Research by Nine Workers' Organisations in Asia and Europe.* Manchester: WWW

Women Working Worldwide (WWW) (2003b) *Subcontracting in the Garment Industry: Workshop Report*. Manchester: WWW

Women Working Worldwide (WWW) (2004) *International Supply Chains and Workers Rights: Seminar and Workshop Report*. Manchester: WWW

Wood A (2001) Value chains: An economist's perspective. *IDS Bulletin* 32(3):41–46

World Trade Organisation (WTO) (2002) 'International Trade Statistics 2002'. HTML file: <URL: www.wto.org/english/res_e/statis_e/its2002_e/chp_4_e.pdf>, last accessed October 2004

Wright M W (2001) A manifesto against femicide. In P Waterman and J Wills *Place, Space, and the New Labour Internationalisms* (pp 246–262). Oxford: Blackwell Publishing

Index

NB *Italic type* for a page locator denotes a box, figure or table.

Printed and bound by CPI Group (UK) Ltd, Croydon, CR0 4YY

27/10/2024

14580367-0004